T0136304

Advances in Computerized Analysis in Clinical and Medical Imaging

Advances in Computerized Analysis in Clinical and Medical Imaging

Edited by

J. Dinesh Peter
Steven Lawrence Fernandes
Carlos Eduardo Thomaz

CRC Press
Taylor & Francis Group
Boca Raton London New York

CRC Press is an imprint of the
Taylor & Francis Group, an **informa** business

CRC Press
Taylor & Francis Group
52 Vanderbilt Avenue,
New York, NY 10017

© 2020 by Taylor & Francis Group, LLC

CRC Press is an imprint of Taylor & Francis Group, an Informa business

No claim to original U.S. Government works

Printed on acid-free paper

International Standard Book Number-13: 978-1-138-33329-1 (Hardback)

Visit the Taylor & Francis Website at
www.taylorandfrancis.com

and the CRC Press Website at
www.crcpress.com

Contents

Preface

In the last decade, a number of sophisticated and new clinical and medical image analysis techniques had been evolved that strides the society in every facet of it. This book provides an ideal reference for all the medical imaging researchers and professionals to explore their innovative methods and analysis on imaging technologies for the better prospective of patient care. The purpose of this compendium is to serve as an exclusive source for the new computer assisted clinical and medical developments in imaging diagnosis, intervention, and analysis. This book will also include articles on computer-assisted medical scanning techniques, computer-aided diagnosis, robotic surgery and imaging, imaging genomics, clinically oriented imaging physics and informatics, augmented-reality medical visualization, imaging modalities, computerized radiology, oncology, and surgery. Moreover, information on nonmedical imaging that has medical applications such as multiphoton microscopy and confocal, photoacoustic imaging, optical microendoscope, infrared radiation, and other imaging modalities would be welcomed.

Each chapter in this book will provide annotations for all the medical and clinical imaging and fundamental advances of clinical and medical image analysis techniques. This book will be a good source for all the medical imaging and clinical research professionals, outstanding scientists, and educators from all around the world for network of knowledge sharing. This book will comprise high-quality disseminations of new ideas, technology focus, research results and discussions on the evolution of clinical and medical image analysis techniques for the benefit of both scientific and industrial developments. This book is devoted to the spreading of knowledge through the publication of scholarly research, primarily in the fields of clinical and medical imaging. The types of chapters consented include those that cover the development and implementation of algorithms and strategies based on the use of geometrical, statistical, physical, functional to solve the following types of problems, using medical image datasets: visualization, feature extraction, segmentation, image-guided surgery and intervention, digital anatomical atlases, representation of pictorial data, statistical shape analysis, computational physiology, virtual and augmented reality for therapy planning and guidance, telemedicine with medical images, telepresence in medicine, and image-guided surgeries.

About the Editors

J. Dinesh Peter is currently working as associate professor, Department of Computer Science and Engineering at Karunya Institute of Technology and Sciences, Coimbatore, Tamil Nadu, India. Prior to this, he was a full time research scholar at National Institute of Technology, Calicut, India, from where he received his Ph.D. in computer science and engineering. His research focus includes Big data, image processing, and computer vision. He has several publications in various reputed international journals and conference papers which are widely referred to. He is a member of IEEE, MICCAI, Computer Society of India, and Institution of Engineers India and has served as session chairs and delivered plenary speeches for various international conferences and workshops. He has conducted many international conferences and been an editor for Springer proceedings and many special issues in journals.

Steven Lawrence Fernandes is currently working as a postdoctoral researcher in the area of deep learning under the guidance of Prof. Sumit Kumar Jha at The University of Central Florida, USA. He also has postdoctoral research experience working at The University of Alabama at Birmingham, USA. He has his Ph.D. in computer vision and machine learning from Karunya Institute of Technology and Sciences, Coimbatore, Tamil Nadu, India. His Ph.D. work "Match Composite Sketch with Drone Images" has received patent notification (Patent Application Number: 2983/CHE/2015) from Government of India, Controller General of Patents, Designs & Trade Marks. He has received the prestigious U.S. award from Society for Design and Process Science for his outstanding service contributions in the year 2017 and Young Scientist Award by Vision Group on Science and Technology, Government of Karnataka, India in the year 2014. He also received research grant from University of Houston Downtown, USA and The Institution of Engineers (India), Kolkata, India. He has collaborated with various scientists, professors, researchers, and jointly published more than 50 research articles which are in Science Citation Indexed (SCI) Journals.

Carlos Eduardo Thomaz holds a degree in electronic engineering from the Pontifical Catholic University of Rio de Janeiro (1993), a master's degree in electrical engineering from the Pontifical Catholic University of Rio de Janeiro (1999), a Ph.D. and a postdoctoral degree in computer science from Imperial College London (2005). He is a full professor at FEI's University Center. He has experience in the area of computer science, with emphasis on pattern recognition in statistics, working mainly in the following subjects: computational vision, computation in medical images, and biometrics.

Contributors

Saravanan Alagarsamy
Assistant Professor
Department of Computer Science and
  Engineering
Kalasalingam Academy of Research and
  Education (KARE)
Tamil Nadu, India

Deepthy Mary Alex
Department of Electronics and Communi-
  cation Engineering
Karunya Institute of Technology and
  Sciences
Coimbatore, Tamil Nadu, India

S. Meenakshi Ammal
Research Scholar, Department of
  Computer Science and Engineering,
  PSG
  College of Technology
Coimbatore, Tamil Nadu, India

Mary X. Anitha
Department of Instrumentation
  Engineering
Karunya Institute of Technology and
  Sciences
Coimbatore, Tamil Nadu, India

S. Arockiaraj
Assistant Professor
Department of Electrical and Electronics
  Engineering
Mepco Schlenk Engineering College
  (Autonomous)
Sivakasi, Tamil Nadu, India

A. Asaithambi
School of Computing
University of North Florida
Jacksonville, FL, USA

J. Joshan Athanesious
Research Scholar
Department of Electronics
  Engineering
Madras Institute of Technology,
  Chromepet, Chennai, India

Caren Babu
Research Scholar
Department of Electronics and
  Communication Engineering
Karunya Institute of Technology and
  Sciences
Coimbatore, Tamil Nadu, India

K. Banumalar
Assistant Professor (Senior Grade)
Department of Electrical and Electronics
  Engineering
Mepco Schlenk Engineering College
  (Autonomous)
Sivakasi, Tamil Nadu, India

Rodrigo P. Bechelli
Department of Electrical Engineering
Av. Humberto de Alencar Castelo Branco
  São Bernardo do Campo
São Paulo, Brazil

Aldo A. Belardi
Department of Electrical Engineering
Av. Humberto de Alencar Castelo Branco
São Bernardo do Campo
São Paulo, Brazil

B. Booba
School of Computing Science
Vels Institute of Science, Technology and
  Advanced Studies (VISTAS)
Pallavarm, Chennai, Tamil Nadu,
  India

D. Abraham Chandy
Department of Electronics and Communi-
cation Engineering
Karunya Institute of Technology and
Sciences
Coimbatore, Tamil Nadu, India

P. Chitra
Research Scholar
Department of Computer Science and
Applications, Gandhigram Rural
Institute (Deemed to be University)
Gandhigram, Dindigul, India

J. Shiny Christobel
Assistant Professor
Department of Electronics and
Communication Engineering
Sri Ramakrishna Institute of
Technology
Coimbatore, Tamil Nadu, India

S. Deivarani
Assistant Professor
Department of Computing
Coimbatore Institute of
Technology
Coimbatore, Tamil Nadu, India

R. Renuga Devi
School of Computing Science
Vels Institute of Science, Technology and
Advanced Studies (VISTAS)
Pallavarm, Chennai, India

D. Raveena Judie Dolly
Assistant Professor
Department of Electronics and
Communication Engineering
Karunya Institute of Technology and
Sciences
Coimbatore, Tamil Nadu, India

G. Emayavaramban
Department of Electric and Electronic
Engineering, Karpagam Academy of
Higher Education
Coimbatore, Tamil Nadu, India

Fernandho de O. Freitas
Department of Electrical Engineering
Av. Humberto de Alencar Castelo
Branco São Bernardo do Campo
São Paulo, Brazil

Xiao-Zhi Gao
School of Computing
University of Eastern Finland
Kuopio, Finland

R. Gayathri
Department of Computing
Coimbatore Institute of Technology
Coimbatore, Tamil Nadu, India

M. Jayesh George
Department of Electronics and
Communication Engineering
Vimal Jyothi Engineering College
Kannur, Kerala, India

Vishnuvarthanan Govindaraj
Associate Professor, Department of
Biomedical Engineering
Kalasalingam Academy of Research and
Education (KARE)
Tamil Nadu, India

Oshin R. Jacob
Department of Computer Science and
Engineering
Karunya Institute of Technology and
Sciences
Coimbatore, Tamil Nadu, India

D. J. Jagannath
Assistant Professor
Department of Electronics and
Communication Engineering
Karunya Institute of Technology and
Sciences
Coimbatore, Tamil Nadu, India

R. Anup Raveen Jaison
Department of Computer Science and
Engineering
SASTRA Deemed University
Thanjavur, Tamil Nadu, India

S. Jayanthy
Department of ECE
Sri Ramakrishna Engineering
  College
Coimbatore, Tamil Nadu,
  India

L. S. Jayashree
Professor
Department of Computer Science and
  Engineering
PSG College of Technology
Coimbatore, Tamil Nadu, India

Kartheeban Kamatchi
Associate Professor
Department of Computer Science and
  Engineering
Kalasalingam Academy of Research and
  Education (KARE)
Tamil Nadu, India

Aldrin Karunaharan
Department of Process Engineering
International Maritime College
Sultanate of Oman

R. Krithiga
Department of Electronics and Instrumen-
  tation Engineering
SRM Valliammai Engineering
  College
Chennai, Tamil Nadu, India

B. V. Manikandan
Senior Professor
Department of Electrical and Electronics
  Engineering
Mepco Schlenk Engineering College
  (Autonomous)
Sivakasi, Tamil Nadu, India

M. Marimuthu
Assistant Professor
Department of Computing
Coimbatore Institute of Technology
Coimbatore, Tamil Nadu, India

S. Naganandhini
Department of Computer Science and
  Applications
Gandhigram Rural Institute (Deemed to
  be University)
Gandhigram, Dindigul, India

C. Narendhar
Department of Nano Sciences
Sri Ramakrishna Engineering College
Coimbatore, Tamil Nadu, India

J. Macklin Abraham Navamani
Department of Computer Applications
Karunya Institute of Technology and
  Sciences
Coimbatore, Tamil Nadu, India

Sundar G. Naveen
Department of Computer Science and
  Engineering
Karunya Institute of Technology and
  Sciences
Coimbatore, Tamil Nadu, India

Felix Erdmann Ott
Data Scientist
Technische Universität Berlin
Germany

Anand Paul
School of Computer Sciences and
  Engineering
Kyungpook National University
South Korea

Antonitta Eileen Pious
Department of Computer Applications
Sri Krishna College of Engineering and
  Technology
Coimbatore, Tamil Nadu, India

Rodrigo G. G. Piva
Department of Electrical Engineering
Av. Humberto de Alencar Castelo Branco
  São Bernardo do Campo
São Paulo, Brazil

A. Prema
School of Computing Science
Vels Institute of Science, Technology and
    Advanced Studies (VISTAS)
Pallavarm, Chennai, Tamil Nadu,
    India

J. Anand Pushparaj
NIT
Tiruchirappalli, Tamil Nadu, India

S. Ramkumar
School of Computing, Kalasalingam
    Academy of Research and
    Education
Krishnankoil, Virudhunagar (Dt)
    Tamil Nadu, India

M. Mary Shanthi Rani
Assistant Professor
Department of Computer Science and
    Applications
Gandhigram Rural Institute (Deemed to
    be University)
Gandhigram, Dindigul, India

M. Mohammed Mansoor Roomi
Department of Electronics and
    Communication Engineering
Thiagarajar College of Engineering
Madurai, Tamil Nadu, India

Lina Rose
Department of Instrumentation Engineering
Karunya Institute of Technology and
    Sciences
Coimbatore, Tamil Nadu, India

S. Perumal Sankar
Department of Electronics and
    Communication Engineering
TocH Institute of Science and
    Technology
Ernakulam, Kerala, India

P. Shanmugavadivu
Department of Computer Science and
    Applications
Gandhigram Rural Institute (Deemed to
    be University)
Gandhigram, Dindigul,
    India

V. Sivakumar
Assistant Professor
Technology Park Malaysia
School of Computing, FCET, APU
BukIt Jalil, Kuala Lumpur,
    Malaysia

R. Sivaranjani
Department of Computer Science and
    Engineering
KPR Institute of Engineering and
    Technology
Coimbatore, Tamil Nadu,
    India

P. Sriramakrishnan
School of Computing, Kalasalingam
    Academy of Research and
    Education
Krishnankoil, Virudhunagar (Dt)
    Tamil Nadu, India

S. Jayanthi Sree
Department of Electronics and
    Communication Engineering
Government College of
    Engineering
Salem, Tamil Nadu, India

D. Sugumar
Department of Electronics and
    Communication Engineering
Karunya Institute of Technology and
    Sciences
Coimbatore, Tamil Nadu, India

S. Sundaramahalingam
Assistant Professor
Department of Electrical and
   Electronics Engineering
Mepco Schlenk Engineering College
   (Autonomous)
Sivakasi, Tamil Nadu, India

Sridevi Unni
Department of Computer Applications
Sri Krishna College of Engineering and
   Technology
Coimbatore, Tamil Nadu, India

M. S. Aezhisai Vallavi
Assistant Professor
Department of Mechanical Engineering
Government College of Technology
Coimbatore, Tamil Nadu, India

P. T. Vanathi
Professor
Department of Electronics and
   Communication Engineering
PSG College of Technology
Coimbatore, Tamil Nadu, India

C. Vasanthanayaki
Department of Electronics and
   Communication Engineering
Government College of Engineering
Salem, Tamil Nadu, India

S. Yogashri
Department of Electronics and
   Communication and Engineering
Sri Ramakrishna Engineering College
Coimbatore, Tamil Nadu, India

N. Yuvaraj
KPR Institute of Engineering and
   Technology
Department of Computer Science and
   Engineering
Coimbatore, Tamil Nadu, India

1

A New Biomarker for Alzheimer's Based on the Hippocampus Image Through the Evaluation of the Surface Charge Distribution

Aldo A. Belardi, Fernandho de O. Freitas, Rodrigo P. Bechelli, and Rodrigo G. G. Piva

Centro Universitário da FEI, Department of Electrical Engineering, Av. Humberto de Alencar Castelo Branco, São Bernardo do Campo, São Paulo, Brazil

1.1 Introduction

Alzheimer's disease (AD) is the most common type of dementia among the elderly population, with no cure and epidemiological trend projections over the next thirty-five years worldwide.[1]

The bibliography shows mild cognitive impairment (MCI)—an AD predecessor phase which a subject undergoes when the initial abnormal functional and structural transformations occur in the brain, given the AD degenerates nerve cells, reducing gray matter volume.[2,3] Some medical imaging and clinical exams can detect these transformations, such as magnetic resonance imaging (MRI) and positron emission tomography (PET). Scientists have struggled to find out that AD and MCI biological markers can help in early disease prediction and adequate treatments. Currently, only a single marker is available that can predict AD.[4] Instead, several groups of biomarkers are used to achieve high sensitivity and specificity for classification among AD, MCI, and normal groups cognition.[4]

Hippocampus volume is the widely used AD biomarker because this is the region of the brain first affected by the disease. There are researches that deeply investigate the hippocampus, aiming to detect transformations due to AD.[5]

In this chapter, we present a novel approach to identify morphological changes in the hippocampus in microstructural level. Based on the concept that electrostatic charges exert influences on each other, we analyze the hippocampus surface charge density distribution (SCDD). The objective is verified if SCDD can provide information about the local and global morphological transformations occurring in the hippocampus. For this purpose, we developed a model to calculate the SCDD and embedded a tool where a user can calculate the hippocampal masks SCDD from images.[6–8]

1.2 Earlier Detection of Alzheimer's Disease

The first known publication of Alzheimer's disease was 100 years ago, but major contributions occurred in the last thirty years, when it started to be treated as lethal and common case of dementia.[9] In 1984, the National Institute of Neurological Disorders and Stroke and the Alzheimer's Association developed the clinical criterion to diagnose Alzheimer's in patients already diagnosed with dementia, according to the Diagnostic and Statistical Manual of Mental Disorders.[10] This method, which we call the "old criteria," was based mainly on neuropsychology, with tests applications to evaluate whether physical and mental disturbances were perceived and associated with characteristic risk factors of AD. Complementary tools were also applied to the old criteria as electroencephalogram (EEG) and computed tomography to detect abnormal structural alterations and brain activities decelerations.[11]

Evaluation by the old criterion did not allow the definitive conclusion of AD with the patient still alive. Instead, definitive diagnosis would only be possible after autopsy, evaluating the presence of amyloid plaques and neurofibrillary brains.

In the 1990s, scientists searched for effective ways to diagnose AD. In the biochemistry field, studies focused in obtaining biomarkers of amyloid by extracting the spinal brain fluid.[12–14]

In the neuroimaging field, studies search for morphological alterations of the brain in MR images correlated with AD, more specifically gray matter (GM). With indication that AD affects several regions of the brain in different ways according to the analyzed group, medial temporal lobe which includes the hippocampus is the region mostly affected with reductions in large vessel.[15,16] MRI scans contributed to the verification of neurofibrillary bundles and GM reduction due to the death of brain cells.

In the late 1990s, MCI was conceptualized as intermediate stage between normal and patient with AD.[17] An individual with MCI may develop AD as well as other types of dementia, stabilized in MCI, or revert to normal control (NC) status. Usually patients with MCI seek specialists because it is at this stage that memory begins to be compromised. It was understood that the study of MCI would be of great importance and one of the main proposals would be to predict whether the patient with MCI would develop AD or not in the future. It would also make it possible to develop drugs to prevent the MCI from developing. In the 2000s, advances in the areas of artificial intelligence and the availability of computers with increasing processing capacities helped researches aimed at the detection of AD, using advanced statistics in the joint analyzes of several biomarkers in the individuals of the populations studied.[18]

At that time, ADNI was created with the purpose of studying how initial MR images and demographic data could be associated with biological markers to study the progression of CL and AD. Currently, ADNI is one of the main databases in the study of the disease.

From a clinical perspective, the current diagnosis of AD is based on the 2011 published version of the "1984 criteria of AD diagnosis." In this version, similar to the old one, diagnosis of individuals with dementia is based on the Diagnostic and Statistical Manual of Mental Disorders 5 (DSM 5) and specialists consult, who may request tests and historical evaluation of the disease.

From a scientific perspective, the last ten years in research focused on the automation of early AD detection, with the patient is still in the state of MCI. Current disease prediction systems correlate information from several biomarkers to achieve high accuracy in early stages.[19]

Structural MR imaging is widely used because of moderate cost and good diagnostic accuracy.[20] The main techniques involve the detection of AD extracting information from the hippocampus, cortex thickness, voxel level probabilities, and texture.

Among the information obtained from the hippocampus, the volume is a widely used marker of AD but may be supplanted by changes in volume in certain regions, compensated by dilations or atrophies in other regions.[21] Also, it supports the analysis of changes in the hippocampus from a microstructural and morphological point of view.[22]

As structural information from the brain are obtained by MR imaging, there are researches that use nuclear medicine technologies to analyze the metabolic behavior of the beta amyloid protein—one of the major biomarkers of AD.[23,24] The limitation of the use of nuclear medicine, such as PET and single-photon emission computed tomography (SPECT) tests, is the unavailability of such equipment and the high cost, which makes it impossible to be used in poor countries.[25] Another disadvantage is the health risk from radioactive drug ingestion, so it would not be advisable to use this technology as the first option when a patient seeks treatment. The advantage of the studies is that the metabolic changes in the brain may precede the structural changes and anticipate earlier diagnosis of AD.

1.3 Database ADNI

The images used in this work were obtained from the ADNI database.[1] ADNI was started in 2003, the public partner company, led by Dr. Michael W. Weiner. The initial objective of ADNI was to test how serial MR images, PET, other biological markers, clinical and neuropsychological assessments could be combined to measure MCI to the progression of early stages of AD. In this database, the MR images are not identified by the names of the individuals, guaranteeing the confidentiality of individuals, and use of images did not require filling forms of Ethics Committee.

The method proposes determining the SCDD in the hippocampus using hippocampal masks as input element images, generated from the MR images and obtained by the ADNI database (orange color in the block diagram of Figure 1.1). The masks have targeted each brain region function between the "hippocampus" and "non-hippocampus." Therefore, they are binary images, where logical values 0 represent the elements that are not hippocampus and 1 represents hippocampus.

Figures 1.2a and 1.2b shows, respectively, the MR image of hippocampus mask (white) and other brain regions (black).

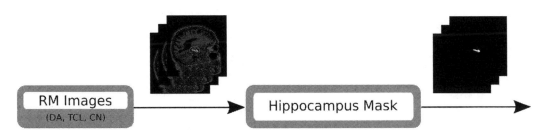

FIGURE 1.1
Block diagram of the images used.

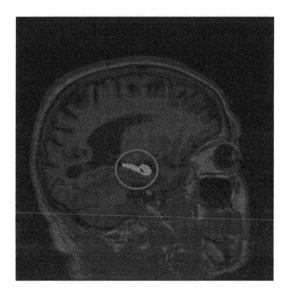

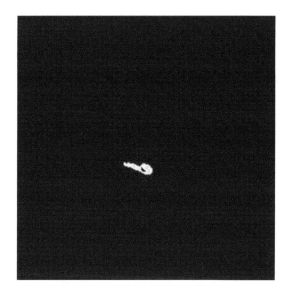

FIGURE 1.2
Example of magnetic resonance imaging (MRI) of the hippocampus (white) and other brain regions (black).

The evaluation tool for performance and sensitivity test and study of electric charge area was used as hippocampus masks in nine individuals—three with DA, three with MCI, and three with NC, totaling 309 sections of the hippocampus processed.

Regarding the studies of SCDD as biomarkers, we compared the hippocampus of a same subject in 2 distinct times. We carried out this observation in one AD subject, 1 MCI and 1 NC, resulting in 6 MRI sections processed.

1.4 Preprocessing

To correct any misalignment and scale differences in the hippocampus, we used six images of the same individual comparisons (comparing SCDD hippocampus) and performed the procedures listed next, using the SPM12 tool.

For the other 309 images (used with sensitivity tests, computational efforts analysis and determination of charges and areas relationships by subregions), whose analyses do not take into account the hippocampus comparisions, the registering process was not necessary.

The two shades of hippocampus of one patient generated in MRI exams on different dates are selected, and in two was taken as spatial alignment reference image whose exam date of MRI is the oldest, A registration routine is executed called "Realign: Estimate & Re-slice" in SPM12 tool available in Matlab.

Due to the interpolation factor in the realignment process, the resultant images are not binary and grayscale. To turn them exclusively into black and white images, the intensity threshold is applied and the method in which it is attributed to black and grayscale tones below it and white above it is determined.

1.5 Region of Interest

The dimensions of the MRI images are approximately 256 × 256 voxels in sagittal plain enough to cover the entire brain structures. The hippocampus masks, generated from MR images, inherit their dimensions. However, a small hippocampus occupies a smaller region in shades. Using lower region of interest (ROI) is preferred from the computational point of view. One reason is that during the SCDD matrix determination

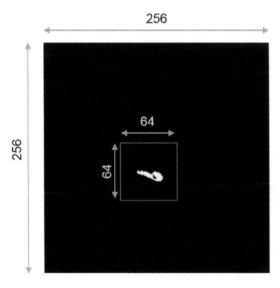

FIGURE 1.3
Example of a region of interest (ROI).

hippocampus masks are generated whose dimensions can reach the dimensions of the second power in high ROI rates. Taking the hippocampal region in the sagittal plane illustrated in Figure 1.3 as an example, an ROI with dimensions equal to 64 × 64 divisions is enough to fully encompass it.

As the dimensions must be powers of two (because of the mathematical function of wavelets), the upper dimensions are immediately 128 × 128. This increase would include more regions that are not hippocampus, marked by black color, and the calculation of SCDD increases the computational effort unnecessarily. In this case, matrices would be generated with dimensions of up to 1282 × 1282, resulting in 268,435,456 instead of 642 × 642 elements, resulting in 16,777,216 elements. The simulations in this work used square ROI with equal sides A64 divisions—dimensions enough to fully cover the hippocampus analyzed.

1.6 Proposed Modeling and Tool

Our tool is represented by the blue part in the block diagram of Figure 1.4. Before we explain about the tool and its application in AD issue, let us begin by the assumption of a hypothetical flat-shaped body, with negligible thickness, permittivity, and electric potential. With all these data, we can find out how the electric charges distribute in this body. We can use Equation (1.1) to calculate surface charge density function denoted as ρ (x, y), which provides the infinitesimal electric charge density in the x and y coordinates. According to this equation, if this flat body changes shape, distribution changes the cartoons too. The more interesting point is that these changes in distribution of charges not only occur in the exact places of shape transformations but also in the whole body. Based on these concepts, we can use the charge distributions to help in the exact identification of body expansion or reduction places and how these changes affect the overall body shape.

Consider that the object is a hippocampus section with the boundary conditions defined in Listing 1. We can compare the distributions with a reference charge in order to monitor the progression of AD, and MCI analysis of this site provides global information about where the hippocampus is expanded or atrophied.

Through the practical point of view, we have developed a tool with optimization features to calculate the charge distributions using matrix notation. The objective of this tool is to determine the matrix of surface charge density distribution (SCDD), analog to ρ (x, y) Equation (1.1). The SCDD is a finite matrix, its elements indicate the surface charge density in small square areas of the object under analysis. All small areas have equal dimensions and are not infinitesimal in ρ (x', y').

Our tool has the flexibility to calculate the SCDD details in different levels. Thus, the researcher can combine different levels of details according to his/her needs and finds the best relationship between detailing and data volume/computational efforts. This is possible because the tool uses Haar wavelet expansion of the functions of ρ (x', y') and takes advantage of their shifting and scaling properties. Haar wavelets also allow calculations with sparse matrices, reducing the computational time during matrix operations.

In a focus tool in the hippocampus, the user inputs the section of hippocampal mask images responsible for providing information about its shape and dimensions voxel, and used in further calculation of distances among charges. The user also has the flexibility to define the detailing levels to calculate the SCDD according to his/her demands. The constants associated with the hippocampus—relative permittivity and electric potential—are already defined in the source code. The tool calculates the internal matrices $[H]$, $[V_m]$, $[V_m']$, $[Z_{mn}]$ $[Z_{mn}']$, $[\rho_n']$, $[\rho_n]$ as shown in Figure 1.4.[26]

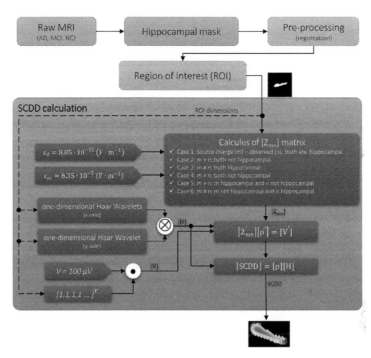

FIGURE 1.4
Tool block diagram with example of application.

LIST OF THE BOUNDARY CONDITIONS CONSIDERING THE HIPPOCAMPUS REGION

The thickness of a flat and uniform surface of a hippocampus section is negligible. The electrostatics equations use double integrals instead of triple integrals in the case of volumes, reducing the computational efforts.

The electrical potential in the hippocampus region is equal to 100 µV, considering this is of the same magnitude found in electroencephalography exams.

The relative permittivity of the hippocampus is equal to the gray matter, as it is the main substance found in this region. This value is $\varepsilon_{GM} = 7.18 \times 10^{-6}$ at low frequencies (70 Hz), gamma within the electroencephalogram band (20–70 Hz).

Considering the hippocampus static charges, the electrostatics equations are applied to the vacuum, given this region has electric potential and current permittivity at low frequencies.

1.7 Mathematical Background

Equation (1.1) from statics shows how the surface electric charge distribution ρ (x, y) creates the electric potential V (x, y)[27]

$$V(x,y) = \frac{1}{4\pi\varepsilon} \int\limits_{-a}^{a} dx' \int\limits_{-b}^{b} \frac{\rho(x',y')}{\sqrt{(x-x')^2 + (y-y')^2}} dy' \tag{1.1}$$

The terms ε, a, and b denote the permittivity, integral axle limits in x and y. The terms $(x-x')$ and $(y-y')$ denote the source and observation distances in each axle. The terms dx and dy denote the paths of integrations.[28–30]

Our objective is to calculate the SCDD numerically using matrix operations and expanding ρ (x', y') into Haar wavelet functions to take advantage of its scaling, shifting, and sparse matrices properties. First, we use the method of moments to expand the linear function ρ (x, y) in a finite series of known function to g_n; the wavelet in our case, with the coefficients α_n, is determined as

$$\rho(x,y) \approx \sum_{n=1}^{N} \alpha_n g_n \tag{1.2}$$

The method of moments involves the use of a weight function, here $W_m = 1$. Applying Equation (1.2) and W_m to Equation (1.1) results in

$$V(x,y) = \langle W_m, f, \mathbf{L}g \rangle$$

$$V(x,y) = \frac{1}{4\pi\varepsilon} \int\limits_{-a}^{a} dx' \int\limits_{-b}^{b} \frac{\sum\limits_{n=1}^{N} \alpha_n g_n(x',y')}{\sqrt{(x-x')^2 + (y-y')^2}} dy' \tag{1.3}$$

Replacing the function g_n by the two-dimensional Haar wavelet expansion of the function, the potential in the surface is defined by[30,31]

$$V(x,y)4\pi\varepsilon = a_j b_j \int\limits_{-a}^{a} \int\limits_{-b}^{b} \frac{\phi(x,y)}{\sqrt{(x-x')^2 + (y-y')^2}} dx\,dy$$

$$+ \left[\sum_{j=-\infty}^{\infty} \sum_{k=-\infty}^{\infty} a_{j,k} b_{j,k} \int\limits_{-a}^{a} \int\limits_{-b}^{b} \frac{\psi_{j,k}^{(H)}(x,y)}{\sqrt{(x-x')^2 + (y-y')^2}} dx\,dy \right] \tag{1.4}$$

where the terms j, k and b_j, k are two-dimensional wavelet coefficients, the j and k indexes are the scaling and wavelet shifting parameters, φ (x, y) the father wavelet scaling function, $\psi_{1,0}$ (x, y) the mother wavelet, and $\psi_{j,k}$ (x, y) the daughter wavelets.

Equation (1.5) shows how to represent Equations (1.3) and (1.4) in matrix notation, where $[Z_{mn}]$ contains the relationship among all charges.[32–34]

$$[V_m] = [Z_{mn}][\alpha_n] \tag{1.5}$$

Depending on the coordinates (x, y), the source and watch cartoons can be in the same place or not. Also, they can be inside or outside the hippocampal mask for the section under analysis, resulting in six conditions. Based on the integral solutions presented, we added four other cases not covered for hippocampal mask section. The elements of $[Z_{mn}]$ are calculated with the following six conditional equations.

The first solution considers that source (m) and observation (n) are the same charges densities $(m = n)$ and inside the hippocampus. The term ε_{GM} denotes the gray matter permittivity, $2b$ denotes the length of side subdivision occupied by n or m[35–38]

$$Z_{mn} = \frac{2b}{\pi \varepsilon_{GM}} ln(1 + \sqrt{2}) \tag{1.6}$$

In the second solution, $m = n$, but are outside the hippocampus region.

$$Z_{mn} = \frac{2b}{\pi \varepsilon_0} ln(1 + \sqrt{2}) \tag{1.7}$$

In the third solution, $n \neq m$, both are inside the hippocampus. ΔS_n denotes the subdivision area occupied by m or n.

$$Z_{mn} = \frac{\Delta S_n}{4\pi \varepsilon_{GM} \sqrt{(x_m - x_n)^2 + (y_m - y_n)^2}} \tag{1.8}$$

In the fourth solution, $n \neq m$, both are outside the hippocampus.

$$Z_{mn} = \frac{\Delta S_n}{4\pi \varepsilon_0 \sqrt{(x_m - x_n)^2 + (y_m - y_n)^2}} \tag{1.9}$$

In the fifth solution, $m \neq n$, with m and n inside and outside the hippocampus, respectively.

$$Z_{mn} = \frac{\Delta S_n}{4\pi \sqrt{\left(\varepsilon_{GM}^2 x_m^2 - 2\varepsilon_{GM}\varepsilon_0 x_m x_n + \varepsilon_0^2 x_n^2\right) + \left(\varepsilon_{GM}^2 y_m^2 - 2\varepsilon_{GM}\varepsilon_0 y_m y_n + \varepsilon_0^2 y_n^2\right)}} \tag{1.10}$$

In the sixth solution, $m \neq n$ with m and n outside and inside the hippocampus, respectively.

$$Z_{mn} = \frac{\Delta S_n}{4\pi \sqrt{\left(\varepsilon_0^2 x_m^2 - 2\varepsilon_0\varepsilon_{GM} x_m x_n + \varepsilon_{GM}^2 x_n^2\right) + \left(\varepsilon_0^2 y_m^2 - 2\varepsilon_0\varepsilon_{GM} y_m y_n + \varepsilon_{GM}^2 y_n^2\right)}} \tag{1.11}$$

These six equations are used to calculate $[Z_{mn}]$ in the tool. As the mask segments the hippocampus regions with different logic values from the other brain regions, the tool can determine if m or n is inside or outside the hippocampus. It can also determine ΔS_n and $2b$.

The tool calculates $[V_m]$ which is the scalar product between the electric potential constant in the source code defined by a column matrix of ones. The number of ones depends on inputted image dimensions.

The tool also calculates the one-dimensional Haar wavelet matrix based on the size of the inputted images and calculates tonsorial product to determine the two-dimensional Haar wavelet. Knowing $[Z_m]$, $[V_m]$, $[H]$ and the detailing levels demanded by the user, the tool calculates the wavelet coefficients and finally the SCDD using the following relations:

$$[Z_{mn}] * [\rho_n] = [V_m] \tag{1.12}$$

$$[Z'_{mn}] * [\rho'_n] = [V'_m] \tag{1.13}$$

else,

$$[Z'_{mn}] = [H][Z_{mn}][H^T] \tag{1.14}$$

$$[\rho'_n] = [H^T]^{-1} * [\rho_n] \tag{1.15}$$

$$[V'_m] = [H][V_m] \tag{1.16}$$

$$[H][Z_{mn}][H^T][H^T]^{-1} * [\rho_n] = [H][V_m] \tag{1.17}$$

Data used in the preparation of this chapter were obtained from the ADNI database (adni.loni.usc.edu). The ADNI was launched in 2003 as a public-private partnership, led by principal investigator Michael W. Weiner, MD. The primary goal of ADNI is to test

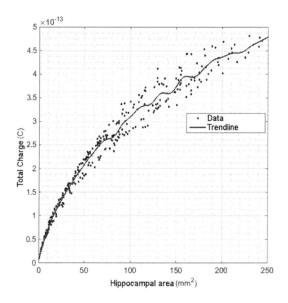

FIGURE 1.5
Total hippocampus electric charge versus its area for each slice tested.

whether MRI, PET, other biological markers, and clinical and neuropsychological assessment can be combined to measure the progression of MCI and early AD. For up-to-date information, visit www.adni-info.org.

1.8 Results

To test the tool, we acquired 30 subjects (10 AD, 10 MCI, 10 NC) hippocampal masks from the ADNI database images. Instead of inputting the hippocampal sections mask directly

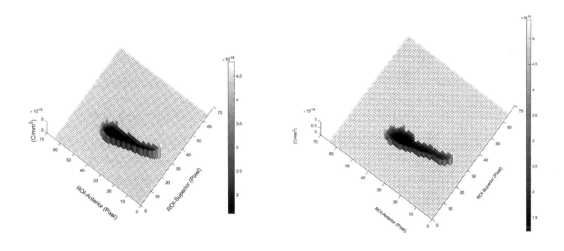

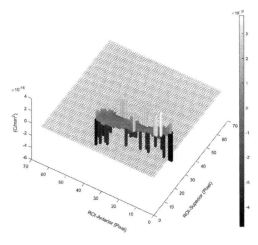

FIGURE 1.6
Different SCDD and their subtraction charts.

to the tool, we extracted smaller ROI composed of 64 × 64 voxels. These dimensions were enough to cover the whole hippocampus tested in sections and reduced unnecessary computational efforts. We used all detailing levels; hence, the dimensions of each SCDD generated were also 64 × 64.

We plotted the SCDD in 2D bar charts, 6.ab, 6.ba, 6.bb, 6.ca, and 6.cb bars denote charge and scale density value by subregions.

We conducted two analysis with the [SCDD] generated. In the first analysis, we calculated the total electric charges in all hippocampus sections and they are verified by the relationship between the total electric charges and their occupied areas. As presented in Figure 1.5, the hippocampus sections with same area but different shapes result in different overall electric charges. This experiment allowed us to verify the influence of the hippocampus in the shape SCDD, as stated by the statics theory.

In the second analysis, we compared two distinct hippocampus SCDD by subtracting each other. 6.ac, 6.bc, and 6.cc results show one subject for each state: AD, MCI, and NC. Checking the subtraction images, we could identify the locations of the hippocampus atrophies (negative bars) and expansions (positive bars), where the density differences are intense. The locations where the charge densities are smaller indicate they were

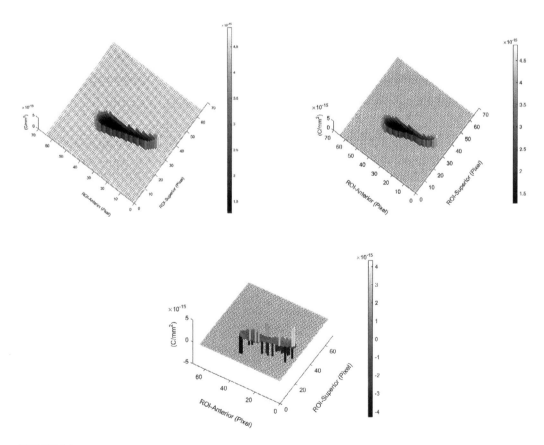

FIGURE 1.6
(Cont.)

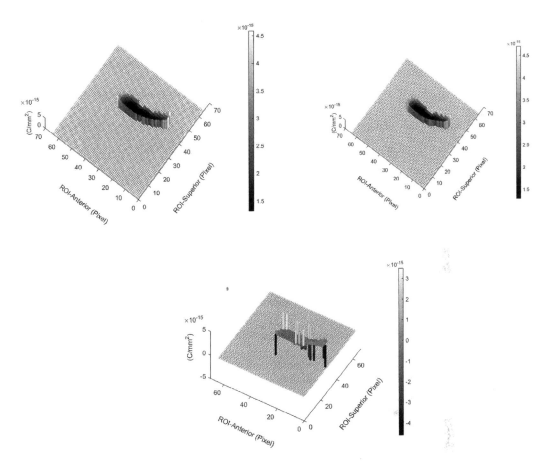

FIGURE 1.6
(Cont.)

transformed not directly, but they were affected by the regions where the morphological changes occurred, mainly if they are close. Therefore, they contribute to recognize the effects of site changes in overall shape of the hippocampus.

Regarding the computational performance aspects, the tool used during the tests generated each SCDD maximum detailing level and plotted them in around 25 s. The computer uses the processor Intel Core i7 2 GHz, with 8 GB RAM and Windows 7 Operational System. Although this time is satisfactory for the author's needs, it is possible to reduce more with optimizations in the source code and improvements in computer resources.

1.9 Conclusion

The mathematical model developed in this chapter is able to calculate SCDD of the hippocampus sections while exploring the Haar wavelet scale and shifting properties that allow users to set the appropriate resolution level in their analysis. We tested the tool in 30 subjects (10 AD, 10 MCI, and 10 NC) hippocampal masks sections acquired from ADNI database. As a result, we could verify the SCDD variations according to the hippocampus shape. The variations of the SCDD not only occur in the exact regions of shape differences but also in the regions with no direct difference, mainly in the adjacent sites. The concepts from electrostatics show each charge exerts influence on each other. Hence, the change in the charge causes changes in the others. By comparing the hippocampus SCDD with a reference, we could identify the exact regions of the hippocampus atrophies and expansions, where the surface charge densities differences are more intense. The regions with smaller changes in SCDD indicate they did not change directly, but according to this magnitude it is possible to know if they are close to regions of few or many direct changes. Hence, they might help to identify the effects of site changes in the overall shape hippocampus.

As a result of this research, SCDD is believed to be a new biomarker for early prediction of AD or in the study of its progression, providing a novel perspective to analyze hippocampal changes through their electrical charges variation. In future work, we intend to find patterns of SCDD in AD, MCI, and NC populations.

References

1. Alzheimer's disease facts and figures. Alzheimer's & Dementia. The Journal of the Alzheimer's Association, p. 332, 2015.
2. Chincarini A., Bosco P., Calvini P., Gemme G., Esposito M., Olivieri C., Rei L., Squarcia S., Rodriguez G., Bellotti R. Local MRI analysis approach in the diagnosis of early and prodromal Alzheimer's disease. Neuroimage, pp. 469–480, 2011.
3. Tondelli M., Wilcock G. K., Nichelli P., De Jager C. A., Jenkinson M., Zamboni G. Structural MRI changes detectable up to ten years before clinical Alzheimer's disease. Neurobiology of Aging, pp. 825–836, 2012.
4. Choi M.-H., Kim H.-S., Gim S.-Y., Kim W.-R., Mun K.-R., Tack G.-R., Lee B., Choi Y. C., Kim H.-J., Hong S. H. Differences in cognitive ability and hippocampal volume between Alzheimer's disease, amnestic mild cognitive impairment, and healthy control groups, and their correlation. Neuroscience Letters, pp. 115–120, 2016.
5. Hua X., Leow A. D., Parikshak N., Lee S., Chiang M.-C., Toga A. W., Jack, Jr., C. R., Weiner M. W., Thompson P. M. Initiative AsDN tensor-based morphometry as a neuroimaging biomarker for Alzheimer's disease: An MRI study of 676 AD, MCI, and normal subjects. Neuroimage, pp. 458–469, 2008.
6. Belardi A. A., Cardoso R. J., Sartori C. A. F. Application of Haar's wavelets in the method of moments to solve electrostatic problems. COMPEL-The International Journal for Computation and Mathematics in Electrical and Electronic Engineering, pp. 606–612, 2004.
7. Belardi A. A., Cardoso J. R., Sartori C. A. F. The Haar wavelets used how expansion function in the method of the moments in the solution of some electrostatic problems. WSEAS Transactions on Mathematics, pp. 603–617, 2010.
8. Harrington, R. F. Field computation by method of moments, NY, Mc Millan, 1968.
9. Alzheimer's Association. Alzheimer's disease facts and figures. Alzheimer's & Dementia, v. 11, 3, pp. 332, 2015.
10. Alzheimer's Association. Changing the Trajectory of Alzheimer's Disease, Chicago, 2015.

11. Mckhann G. Clinical diagnosis of Alzheimer's disease report of the NINCDS-ADRDA Work Group under the auspices of Department of Health and Human Services Task Force on Alzheimer's Disease. Neurology, AAN Enterprises, v. 34, 7, pp. 939–944, 1984.

12. Blennow K. et al. Clinical utility of cerebrospinal fluid biomarkers in the diagnosis of early Alzheimer's disease. Alzheimer's & Dementia, Elsevier, v. 11, 1, pp. 58–69, 2015.

13. Moradi E. Machine learning framework for early MRI-based Alzheimer's conversion prediction in MCI subjects. Neuroimage, Elsevier, v. 104, 398–412, 2015.

14. Molinuevo J. L. et al. The clinical use of cerebrospinal fluid biomarker testing for Alzheimer's disease diagnosis: A consensus paper from the Alzheimer's biomarkers standardization initiative. Alzheimer's & Dementia, Elsevier, v. 10, 6, pp. 808–817, 2014.

15. Chincarini A. et al. Local MRI analysis approach in the diagnosis of early and prodromal Alzheimer's disease. Neuroimage, Elsevier, v. 58, 2, pp. 469–480, 2011.

16. Tondelli, M. et al. Structural MRI changes detectable up to ten years before clinical Alzheimer's disease. Neurobiology of Aging, Elsevier, v. 33, 4, pp. 825, 2012.

17. Petersen R. C. et al. Mild cognitive impairment: Clinical characterization and outcome. Archives of Neurology, American Medical Association, v. 56, 3, pp. 303–308, 1999.

18. Cuingnet R. et al. Automatic classification of patients with Alzheimer's disease from structural MRI: A comparison of ten methods using the ADNI database. Neuroimage, Elsevier, v. 56, 2, pp. 766–781, 2011.

19. Iftikhar M. A. and Idris A. An ensemble classification approach for automated diagnosis of Alzheimer's disease and mild cognitive impairment. In Open Source Systems and Technologies, Institute of Electrical and Electronics Engineers (IEEE), pp. 78–83, 2016.

20. Tong T. et al. A novel grading biomarker for the prediction of conversion from mild cognitive impairment to Alzheimer's disease. IEEE Transactions on Biomedical Engineering, Institute of Electrical and Electronics Engineers (IEEE), v. 64, 1, pp. 155–165, 2017.

21. Lee P. et al. Morphological and microstructural changes of the hippocampus in early MCI: A study utilizing the Alzheimer's disease neuroimaging initiative database. Journal of Clinical Neurology, v. 13, 2017.

22. Frisoni G. B. et al. Mapping local hippocampal changes in Alzheimer's disease and normal ageing with MRI at 3 Tesla. Brain, Oxford University Press, v. 131, 12, pp. 3266–3276, 2008.

23. Bocchetta M. et al. The use of biomarkers for the etiologic diagnosis of MCI in Europe: An EADC survey. Alzheimer's & Dementia, Elsevier, v. 11, 2, pp. 195–206, 2015.

24. Ciblis A. et al. Neuroimaging referral for dementia diagnosis: The specialist's perspective in Ireland. Alzheimer's & Dementia: Diagnosis, Assessment & Disease Monitoring, Elsevier, v. 1, 1, pp. 41–47, 2015.

25. Laske C. et al. Innovative diagnostic tools for early detection of Alzheimer's disease. Alzheimer's & Dementia, Elsevier, v. 11, 5, pp. 561–578, 2015.

26. Datta B. N. Numerical Linear Algebra and Applications, 1st ed., New York, Brooks/Cole Publishing Company, 1995.

27. Constantine A. B. Advanced Engineering Electromagnetics, 2nd ed., New York, John Wiley & Sons, 1989.

28. Harrington, R. F. Field Computation by Moment Methods, 1st ed., New York, Macmillan Company, 1968.

29. Belardi, A. A. et al. Application of Haar's wavelets in the method of moments to solve electrostatic problems. COMPEL: The International Journal for Computation and Mathematics in Electrical and Electronic Engineering, Emerald, v. 23, 3, pp. 606–612, 2004.

30. Belardi A. A., Piccinini A. H. Mathematical modeling for determination the surface charge density and eddy current problem using the Haar wavelet. Journal of Electrical Engineering, David Publishing Company, v. 3, 2, 2015.

31. Morettin P. A. Ondas e ondaletas: da análise de Fourier à análise de ondaletas, São Paulo, Edusp, 1999.

32. Aboufadel E. Schlicker S. Discovering Wavelets, 1st ed., New York, John Wiley & Sons, Inc., 1999.

33. Wojtaszczyk P. A Mathematical Introduction to Wavelets, 2nd ed., New York, Cambridge University Press, 1999.
34. Florkowski M. Wavelet based partial discharge image de-noising. ABB Corporate Research, Poland, v. 5, 8, pp. 21–24, 2000.
35. Antonini G. Orlandi A. Fast Iterative Solution for the Wavelet University of L'Aquila, Italy, 2001.
36. Rabello T. N. Wavelets e redes neurais, V Escola de redes neurais, São José dos Campos, pp. 1–27, 1999.
37. Liang J., Elangovan S., Devotta J. B. X. A wavelet multiresolution analysis approach to fault detection and classification in transmission lines. Elsevier Computer Physics Communications, pp. 327–332, 1999.
38. Unser M., Aldroubi A. A review of wavelets in biomedical application. IEEE Proceedings of the IEEE, v. 84, 4, pp. 626–638, 1996.

2

Independent Vector Analysis of Non-Negative Image Mixture Model for Clinical Image Separation

D. Sugumar

Karunya Institute of Technology & Sciences, Coimbatore, Tamil Nadu, India

P. T. Vanathi

PSG College of Technology, Coimbatore, Tamil Nadu, India

Xiao-Zhi Gao

University of Eastern Finland, Joensuu, Kuopio, Finland

Felix Erdmann Ott

Technische Universität Berlin, Berlin, Germany

M. S. Aezhisai Vallavi

Government College of Technology, Coimbatore, Tamil Nadu, India

2.1 Background and Driving Forces

Lungs cancer is prominent in the world due to various environmental conditions and smoking habits. In India, the death rate is about 6.5% per year and it was reported that there are approximately 320,000 new cases every day in the developing countries. The first step of the treatment starts with the diagnosis and it is done with chest X-ray. If there is a suspect, then a CT scan is recommended. Hence, chest X-ray plays an important role in diagnosing the nodules in the lungs. It would be better if the chest X-ray is processed in such a way that the doctors can diagnose easily. So, the CXR is separated into the bones and the lungs image. Dual-energy subtraction (DES) (Manji et al. 2016)is an advanced technique which is used in hospitals. This technique is static and depends on the fixed coefficients. In this chapter, independent vector analysis (IVA) is applied to separate the bones and lungs in the CXR image which is a dynamic approach.

Cardiovascular diseases include lung cancer, tuberculosis (TB), osteoporosis, etc. The first treatment procedure includes postero anterior (PA) radiography which is nothing but the conventional CXR. This procedure often involves a view from the back to the front of the chest as well as a view from the side. Like any X-ray procedure, chest X-rays

(a)   (b)

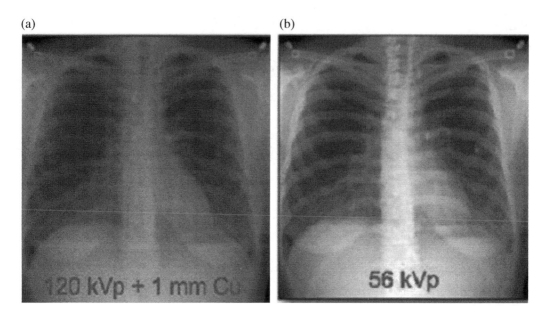

FIGURE 2.1
Dual-energy chest X-ray images: a) high-energy image and b) low-energy image.

expose the patient briefly to a small amount of radiation (Noharaa et al. 2009). The doctors diagnose based on the CXR. If there is a suspect, then the patient is asked to undergo CT scan to further proceed with the treatment. But often the conventional CXR is less sensitive in detecting any nodules. Since detecting and categorizing pulmonary nodules are primary tasks for the chest radiologist, the CXR has to be presented clearly so that they can diagnose correctly.

Dual-energy chest X-ray is an advanced technique of radiography as shown in Figure 2.1. In the dual-exposure technique, high-energy (120 kV$_p$) and low-energy (56 kV$_p$) exposure is used to generate a subtracted bone and a subtracted soft tissue image (Rebuffel & Dinten 2007, Hwang et al. 2008, Kashani et al. 2009). In effect, the contrast of the bone (ribs and spine) was eliminated from the soft tissue image, and the contrast of the soft tissue was greatly reduced in the bone image. By using image processing tools and with the help of processor, the CXR has been separated into bones and lungs images. The technique has been used for more than 15 years and is very useful in finding any cancerous nodules. DES uses some fixed static coefficients to separate the images. Hence, in order to make the separation dynamic, blind signal separation (BSS) technique is used.

2.2 Diseases Diagnosed Using CXR

A chest X-ray is an X-ray of the chest, lungs, heart, large arteries, ribs, and diaphragm. Hence, using the chest X-ray, many diseases can be diagnosed at primary level.

2.2.1 Tuberculosis

Tuberculosis is a common, sometimes fatal infectious disease caused by various strains of mycobacteria, usually *Mycobacterium tuberculosis*. Tuberculosis attacks the lungs and also affects other parts of the body. Tuberculosis creates cavities visible in X-rays. Abnormalities on chest radiographs may be suggestive of, but are never diagnostic of, TB. Pulmonary tuberculosis CXR image is shown in Figure 2.2.

2.2.2 Pneumonia

Pneumonia is a breathing (respiratory) condition in which there is an infection of the lung. It is usually caused by viruses or bacteria and less commonly other microorganisms, certain drugs, and other conditions such as autoimmune diseases. A chest radiograph is usually used for diagnosis. X-ray presentations of pneumonia may be classified as lobar pneumonia, bronchopneumonia, and interstitial pneumonia. Pneumonia-diagnosed CXR image is given in Figure 2.3.

2.2.3 Lung Cancer

Lung cancer is the uncontrolled growth of abnormal cells in one or both lungs. These abnormal cells do not carry out the functions of normal lung cells and do not develop into healthy lung tissue. The chest X-ray is the most common first diagnostic step when any new symptoms of lung cancer are present. Chest X-rays may reveal suspicious areas in the lungs but are unable to determine if these areas are cancerous. In particular, calcified nodules in the lungs or benign tumors may be identified on a chest X-ray. Chest X-ray of lung cancer is shown in Figure 2.4.

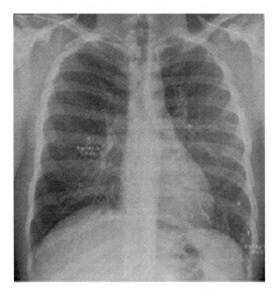

FIGURE 2.2
Chest X-ray of primary tuberculosis. Contributed by Basem Abbas Al Ubaidi.

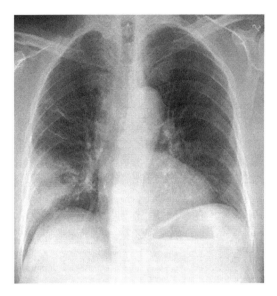

FIGURE 2.3
Chest X-ray of lobar pneumonia. Contributed by Dr. Mikael Häggström.

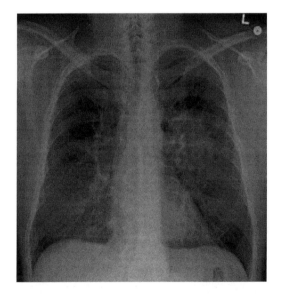

FIGURE 2.4
Lung cancer seen on chest X-ray. Contributed by Dr. James Heilman, MD.

2.2.4 Congestive Heart Failure

Congestive heart failure (CHF) is one of the most common abnormalities evaluated by CXR. CHF occurs when the heart fails to maintain adequate forward flow. CHF may progress to pulmonary venous hypertension and pulmonary edema with

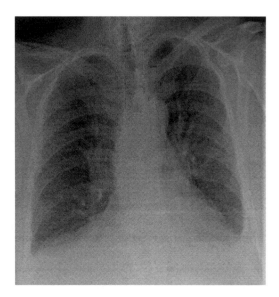

FIGURE 2.5
Congestive heart failure with small bilateral effusions. Contributed by Dr. James Heilman, MD.

leakage of fluid into the interstitium, alveoli, and pleural space. CXR is important in evaluating patients with CHF for development of pulmonary edema and evaluating response to therapy as well. CHF CXR with small bilateral effusions is shown in Figure 2.5.

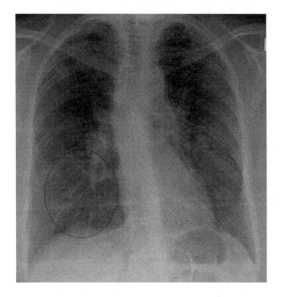

FIGURE 2.6
Chest X-ray showing the typical nodularity of sarcoidosis. Contributed by Dr. James Heilman, MD.

2.2.5 Sarcoidosis

It is a syndrome involving abnormal collections of chronic inflammatory cells (granulomas) that can form as nodules in multiple organs. The granulomas are most often located in the lungs or the lymph nodes. Typical nodularity of sarcoidosis CXR image is shown in Figure 2.6.

2.3 Non-Negative Blind Image Separation

This chapter focuses on developing a multichannel source separation algorithm like IVA (Lee et al. 2007) and its variants (Anderson et al. 2012, Itahashi et al. 2012, Zhang et al. 2012) for non-negative blind signal separation (nBSS). Generally, many real-world data are non-negative such as image, video, and medical data. In fact, the growth of multichannel (array sensors) and multimodel sensors prompted attraction in methods for processing of multi-dimensional and multivariate data. In nBSS, the source images and the mixing matrices are considered to take on non-negative values. In addition, the mixtures are considered as multidimensional and multivariate data. However, the separation of observed mixed images is imperative, since the number of components mixed and the pattern of mixing are unknown. With these unknown parameters, blind image separation (BIS) plays a critical role in the separation of non-negative image mixtures. Hence, nBSS (or BIS) is the main focus of this chapter considering the IVA and its variant for application of image analysis.

BSS or BIS is better understood by means of cocktail party problem (Comon 1994, Hyvarinen et al. 2001, Comon & Jutten 2009). The challenge in BSS is that it is unaware of the number of mixed sources and the pattern of mixing. This criterion also applies to images where acquired images in most cases are mixed. Separation of image mixtures finds its application in many real-time cases like overlapped fingerprint images, reflected images (Bronstein et al. 2005), hyperspectral images (Wang & Chang 2006, Zymnis et al. 2007), bleed-through images (Tonazzini et al. 2006), biomedical images (Wang et al. 2006, Chan et al. 2008), scanned document images (Merrikh Bayat et al. 2011), show-through images (Almeida & Almeida 2012), Dual-Energy X-Ray Image Decomposition (Suzuki et al. 2004, Wang et al. 2010, Chen et al. 2014) and so on. In recent years, the separation of these images is a great deal and the separated images provide a wide range of information compared to the mixture. There are different techniques of BIS for the separation of image mixture which differ from each other by its algorithm and its assumptions.

Oja and Plumbley (2003) discussed about the alternative way of approaching the ICA problem with the additional assumption of non-negativity between the source components to the usual assumption of independent non-Gaussian sources and full rank mixing matrix. Jenssen et al. (2006) designed the nonparametric ICA, which is based on Parzen window. A new convex analysis framework for nBSS, known as CAMNS, is designed by Chan et al. (2008). Arai (2012) developed image source separation based on ICA in wavelet domain. Contrast function is used to determine the Gaussianity between the pixels because non-Gaussian vectors are the independent vectors. The divergence measures are used to determine the dependency or independency between the estimated source components in ICA procedure (Cichocki & Amari 2010). The KL divergence is measured for the joint PDF and the marginal PDF of the individual components. In the same way, α divergence and convex divergences are measured using the PDF of the individual random variables. These divergence measures should be minimized for increased independency. However, C-ICA suffers from the problems of slow convergence rate and the lack of precision when it is applied to multidimensional and multivariate data. Hence, in this chapter pair of novel convex functions has been suggested to overcome these problems. C-ICA is applied to audio signals and has achieved faster separation by Chien and Hsieh (2012). Apart from the general convex function, the special functions like convex-Shannon and convex-logarithm functions were also adopted by the author as contrast function. In line with these convex functions, in this chapter, 2D convex functions are presented.

The purpose of this chapter is to design convex contrast functions and incorporate them into the procedures of CDIV-IPA (non-negative assumption). Hence, the CDIV-IPA approaches are flexible with a controllable convexity parameter. Therefore, it permits the designer to toggle between the degree of convexity and statistical quality. Moreover, the proposed contrast functions are globally convex; therefore, it is possible to achieve numerical stability. The convexity of the contrast functions is of premier importance in many source separation applications as it provides a faster convergence and has an ample range of the gradient descents to reach its minima points. As a consequence, in this chapter, statistically quality separation at a faster convergence rate was obtained by employing effective convex contrast functions (Cosh & LoG). These developed convex contrast functions were primarily enforced to the pixel level row vector or column vector of image mixtures in order to achieve better separation. Experiments on the BIS of instantaneous mixture further demonstrate the high quality (faster convergence and efficient estimate) of the proposed optimized CDIV-IPA to traditional FastICA, CAMNS, and C-ICA algorithms.

2.4 Mixture of Non-Negative Sources Model for Image Separation Task

The general block diagram of blind image separation is shown in Figure 2.7. The problem of blind image separation (i.e., BIS task) is exemplified with mixture of non-negative (MoN) sources model and its mathematical implication. It is assumed that there are N non-negative sources and M non-negative observed mixtures. In the multichannel BIS setting, the observed non-negative mixtures are composed of image mixtures $X_+ = \{x_1, x_2, x_3, \ldots, x_M\}$. Each image mixture is a size of p × p or a single dimensional signal (if the profile of the image is considered) with a sample size of $1 \times p^2$ (row vector size) or $p^2 \times 1$ (column vector size). Basically, the observed non-negative data $X \in R_+^n$: n-dimensional non-negative real numbers set (i.e., $x_i \geq 0$ for $i = 1, 2, 3, \ldots, M$ or $X \geq 0$) is the classical instantaneous linear combination of N non-negative source images S. These non-negative source images $S_+ = \{s_1, s_2, s_3, \ldots, s_N\} \in R_+^n$ (i.e., $s_j \geq 0$ for $j = 1, 2, 3, \ldots, N$ or $S \geq 0$) have a size of q × q which represents the 2D image or a single dimensional sample size of $1 \times q^2$ (row vector size) or $q^2 \times 1$ (column vector size) which represents a vectorized source image. Therefore, MoN sources model is expressed in vector representation as shown in (2.1) and shown in Figure 2.8

$$X_+ = A_+ S_+ \tag{2.1}$$

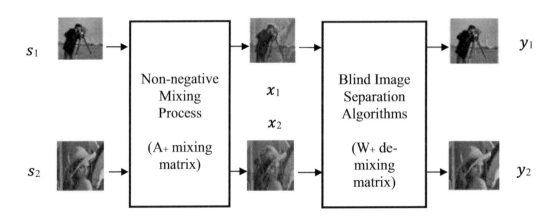

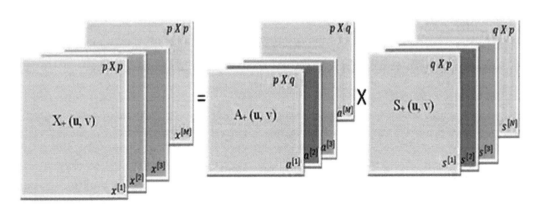

FIGURE 2.7
Block diagram of determined blind image separation.

FIGURE 2.8
Mixture of non-negative sources (MoN) model.

where A_+ denotes the mixing coefficient matrix and it must be invertible, X_+ is observed image mixture, and S_+ is an unknown source image matrix.

The separated images are estimated from the following equation:

$$Y_+ = W_+X_+ \tag{2.2}$$

where W_+ is a demixing matrix and the inverse of the mixing matrix A_+. Y_+ is the estimated image matrix. Consider that the number of sources and the observations are same, then the problem of separation is brought down to a determined BIS case (i.e., $N = M$). The matrix representation of the image to be estimated is represented in M sources and M observations. The coefficients of the mixing matrix (or the demixing matrix datasets) are varying depending on the mixing process of the sources. The mixing process is modeled either linear or nonlinear. Examples of the mixing process in the case of the linear mixing model are: (1) semireflection (glassy window) in superimposed image mixture, (2) merging the shifted version of the same image in ghosting image mixture, (3) defocus blurring (tomography and microscopy images) in convolutive image mixture, and (4) temporal patterns in multichannel biomedical image mixture. Hence, it is difficult to extract the sources with the general BSS mixture model. However, a little knowledge on mixing matrix datasets or healthy assumption on mixing matrix datasets will make the task easier and brings down the general BSS mixture model into a specific BSS mixture model. For multichannel image mixture, the BIS framework is suggested to hold the following assumptions:

(i) Non-negative source images: All source images $S_+ = \{s_1, s_2, s_3, ..., s_M\}$ are non-negative and each and every pixel of the source image are non-negative, $S_+ \in R_+^n$ (i.e., $S \geq 0$ or $s_j(u, v) \geq 0$ for $j = 1, 2, 3, ..., M$ and $u, v = 1, 2, 3, ..., q$), where u and v are the spatial coordinates of the image.

(ii) Non-negative mixing matrix: All mixing matrices $A_+ = \{a_1, a_2, a_3, ..., a_N\}$ are non-negative and all the coefficients of the mixing matrix are non-negative $A_+ \in R_+^{MXM}$ (i.e., $A \geq 0$ or $a_i(u, v) \geq 0$ for $i = 1, 2, 3, ..., M$).

(iii) Independent mixing matrix: If the number of observed mixtures M is greater than or equal to the number of non-negative sources N (i.e., $M \geq N$), then non-negative mixing matrix A or each mixing matrix $a_1, a_2, a_3, ..., a_M$ is linearly independent.

(iv) Additivity of mixing matrix: The sum of all the elements of each row (column) vector is unity.

Assumptions (i) and (ii) are valid in multidimensional data analysis like image, video, and medical data analysis where the linear mixing model is defined by non-negative real-world sources and the non-negative mixing matrix. Assumptions (i) and (ii) are applicable to BIS, because all the observed images are non-negative. Hence, for the MoN model, assumptions (i) and (ii) are the strongest candidates. Assumption (iii) is an assumption generally applicable to BSS and comparatively reasonable to BIS, because the observed mixture images are correlated or dependent on multichannel biomedical image analysis. Assumptions (iii) and (iv) are crucial for the development of BIS or nBSS model. In spite of considerable advancement in various nBSS techniques, accurate separation with a better convergence rate and stability of correlated or dependent sources still remains a challenging task. By considering these four assumptions, the main focus of this chapter is to design CDIV-IPA (non-negative assumption) to achieve accurate separation.

2.4.1 2D Convex Functions

The convex divergence measure is formed using convex function and Jensen's inequality. Jensen's inequality: If φ is convex, then for $0 \leq \leq 1$,

$$\varphi(\lambda x + (1 - \lambda)y) \leq \lambda\varphi(x) + (1 - \lambda)\varphi(y) \tag{2.3}$$

φ is strictly convex if dom is convex for $x, y \in dom\varphi$, $0 < \lambda < 1$. The proposed 2D strong convex function (Cosh) and high convex or strictly convex function (LoG-Laplacian of Gaussian) have met the above conditions. They are given by the following equations:

$$\varphi_{\text{SC/HC}}(u,v) = \frac{\sigma^4}{\pi}\left(e^{-\frac{(u^2+v^2)}{2\sigma^2}}\cosh\left(\frac{(u^2+v^2)}{2}\right)^2 - 1\right) \tag{2.4}$$

$$\varphi_{\text{High/LoG}}(u,v) = \frac{\sigma^4}{\pi}\left(e^{-\frac{(u^2+v^2)}{2\sigma^2}}\left(\left[\frac{(u^2+v^2)\sigma^2}{2}\right] - 1\right)\right) \tag{2.5}$$

where u and v are spatial coordinates and σ is convexity parameter. Additionally, the Cosh function has set of minima point at the bottom of the bowl, hence, it is named strong convex function. At the same time, the LoG function has a higher rate of roll-off curves, hence it is called high convex function.

2.4.2 Divergence Measure

The solution of Chien's work was used and applied to the proposed strongly convex Cosh and highly convex LoG. Doing this new divergence measurements were created with a lower bound of σ (equals to 2.4). Incorporating joint distribution ($jp = P(x_1, x_2)$) and marginal distribution ($mp = (P(x_1)P(x_2))$) into Cosh convex function and applying them to Jensen's inequality yields

$$D_{JSC}(P(x_1,x_2), P(x_1)P(x_2), \sigma) = D_{\text{Cosh}}(jp, mp, \sigma) = \lambda\text{Cosh}(jp) + (1-\lambda)\text{Cosh}(mp)$$
$$- \text{Cosh}(\lambda jp + (1-\lambda)mp) \tag{2.6}$$

Similarly, for LoG convex function, the convex divergence measure is given by

$$D_{JSC}(P(x_1,x_2), P(x_1)P(x_2), \sigma) = D_{\text{LoG}}(jp, mp, \sigma) = \lambda\text{LoG}(jp) + (1-\lambda)\text{LoG}(mp)$$
$$- \text{LoG}(\lambda jp + (1-\lambda)mp) \tag{2.7}$$

2.5 Convex Divergence IPA Framework

The demixing matrix W is estimated by the second-order gradient descent algorithm,
   Column wise update:

$$W(u+1,v) := W(u+1,v) - \eta\frac{\partial^2 D_c(X, W(u+1,v))}{\partial W(u+1,v)[W(u+1,v)]^T} \tag{2.8}$$

Row wise update:

$$W(u,v+1) := W(u,v+1) - \eta\frac{\partial^2 D_c(X, W(u,v+1))}{\partial W(u,v+1)[W(u,v+1)]^T} \tag{2.9}$$

where u,v are the iteration indexes, η is the learning rate or step size, X is the mixture image, and W is the demixing matrix. The divergence measure with respect to individual component in the demixing matrix is described in the second term (gradient term)

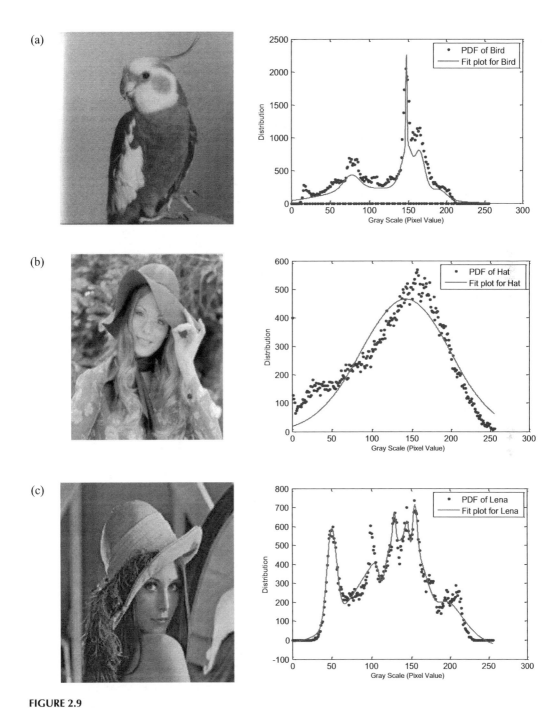

FIGURE 2.9

Standard images: (a) bird and its distribution, (b) hat and its distribution, (c) Lena and its distribution, (d) baboon and its distribution, (e) cameraman and its distribution, and (f) onion and its distribution.

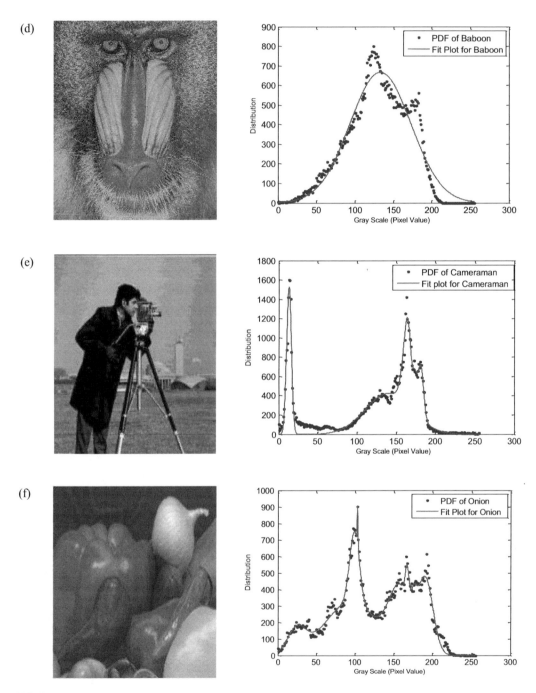

FIGURE 2.9
(Cont.)

of the weight update equations (2.8) (column wise) and (2.9) (row wise) and the gradient of the divergence measure is calculated by the following equation:

$$D_C(X, W, \sigma) = \int\int \left(\lambda\varphi(p(W, X)) + (1-\lambda)\varphi(p(W)p(X)) - \varphi(\lambda p(W, X))\right.$$
$$\left. + (1-\lambda)p(W)p(X)\right)dXdW \qquad (2.10)$$

where σ is the convex parameter of Cosh and LoG.$p(W)$, $p(X)$ are the individual PDFs and $p(W, X)$ is the joint PDF. The SIR is expressed by

$$SIR = 10log_{10}\frac{\sum_{n=0}^{N-1}\hat{S}(n)^2}{\sum_{n=0}^{N-1}\left(S(n) - \hat{S}(n)\right)^2} \qquad (2.11)$$

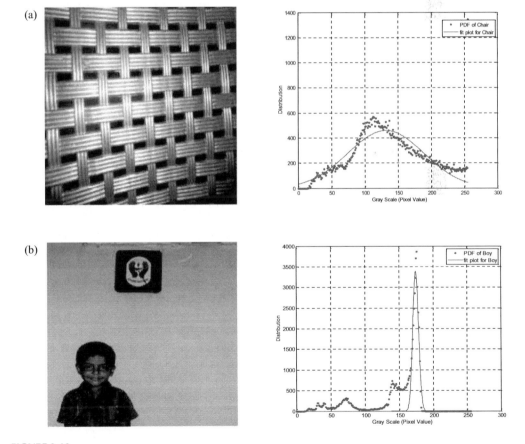

FIGURE 2.10
Mixture of boy and chair image: a) chair and its distribution, b) boy and its distribution, c) mixed images, and d) mixtures distribution.

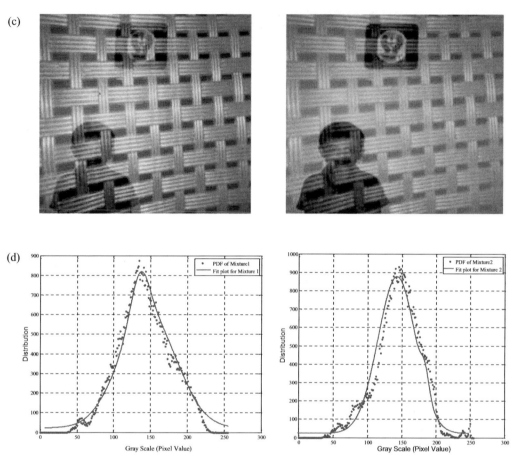

FIGURE 2.10
(Cont.)

where $\hat{S}$ is estimated (target) signal, S is true source signal, and N is the number of samples. The difference of S and $\hat{S}$ is the interference.

2.6 Experimental Setups

The database utilized for analysis comprises of six standard images and two test images of size 256×256 with grayscale ranging from 0 to 255. Randomly ten combinations were considered to test whether the algorithms effectively separated the mixed images. Hat, Lena, baboon, and onion images have a skew value less than 1; hence, they all have less skewed distribution. Whereas, bird, cameraman, boy, and chair images have a skew value greater than 1, so they have heavily skewed distribution. Similarly, the peak shapes are determined by kurtosis value. Bird, cameraman, and boy images have the highest kurtosis value peaked distribution. However, the sources histogram (PDF) plots are tested with a Gaussian curve fitting which also conveyed the same regarding the Gaussianity. The standard images and their curve fitting of PDF are shown in Figure 2.9. The test images, their mixtures and the corresponding curve fitting of PDF are shown in Figure 2.10.

2.6.1 Experiment 1: CDIV-IPA with Non-Negative Assumption

In this experiment, noise-free ten image mixtures of size 256 × 256 were taken with non-negative assumption. The proposed algorithm CDIV-IPA is applied on all the mixtures and the separation quality is measured by SIR which is tabulated for both Cosh and LoG functions in Table 2.1. It is derived from the table that if the PDF of both the images are similar or closer, then the proposed algorithm yields a lesser SIR or lesser separation quality compared to images mixed with different distributions. The BIS task of "hat and baboon," "hat and chair," and "chair and baboon" mixtures yielded SIR less than 20 dB, resulting in poor quality of separation. Only exceptional cases of this analysis were obtained for CDIV-IPA (Cosh) during the separation of Lena and cameraman images as the SIRs obtained are 18.32 dB and 15.54 dB, respectively. The reason for this poor performance is that both images have dominant multimodal distribution.

Randomly selected ten mixtures are used to evaluate the performance of the CDIV-IVA algorithm (Cosh and LoG), and three reported BIS algorithms Fast ICA, CAMNS, and C-ICA. One typical mixture (test images) among the ten is shown in Figure 2.11. It is observed from Figure 2.11 that Fast ICA does not offer accurate separations. CAMNS and C-ICA provide moderate separations and the best results are obtained by CDIV-IPA (Cosh) and CDIV-IPA (LoG). The SIR comparison of all the five algorithms is summarized in bar chart shown in Figure 2.12. It indicates that the LoG-based CDIV-IPA outperformed the other reported algorithms.

TABLE 2.1

SIR (dB) values of the recovered sources by CDIV-IPA

Combination of standard images	CDIV-IPA (cosh)		CDIV-IPA (LoG)	
	SIR (dB) of recovered Image 1	SIR (dB) of recovered Image 2	SIR (dB) of recovered Image 1	SIR (dB) of recovered Image 2
Bird and hat	29.68	29.80	29.73	29.85
Hat and baboon	13.11	10.37	19.14	18.41
Hat and onion	28.42	29.56	29.43	27.57
Hat and chair	17.48	18.20	17.60	18.31
Lena and baboon	28.43	28.60	28.44	28.62
Lena and cameraman	18.32	15.54	21.33	19.57
Baboon and onion	29.51	29.83	29.63	29.89
Baboon and chair	18.72	18.81	18.79	18.90
Cameraman and onion	29.33	30.58	31.40	32.62
Chair and boy	31.25	31.23	33.12	34.87

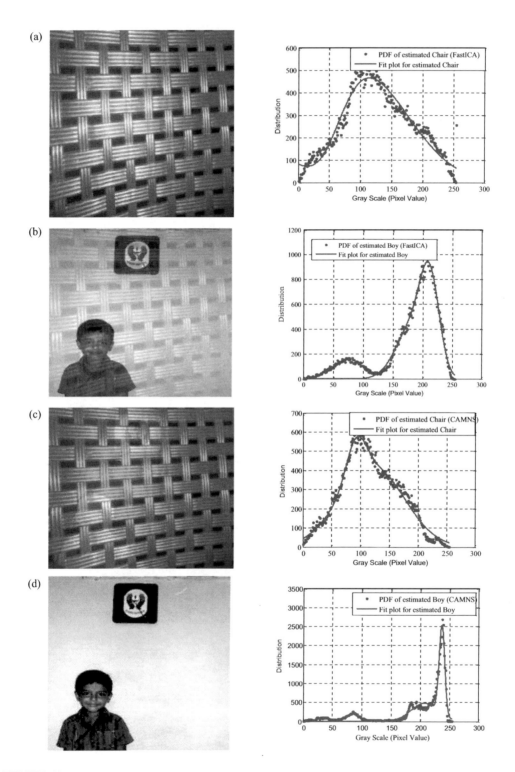

FIGURE 2.11
Separated images and their distribution for the mixture of boy and chair image of different algorithms: a) & b) FAST ICA, c) & d) CAMNS, e) & f) C-ICA, g) & h) CDIV-IPA (COSH), and i) & j) CDIV-IPA (LoG).

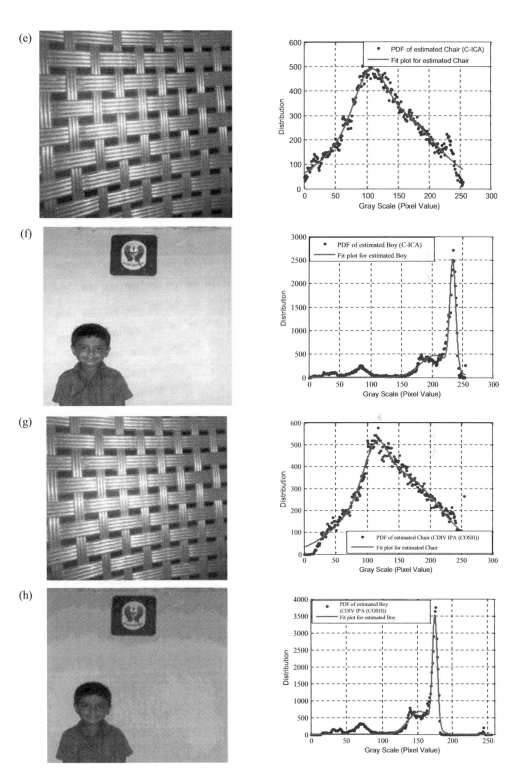

FIGURE 2.11
(Cont.)

2.6.2 Experiment 2: Clinical Image Data

In CXR imaging for precise detection of lung diseases, the soft tissue image which needs to be separated from the mixed image is obtained by the difference of a bone image from the original image. However, the bone images are not available directly for few cases. In computed tomography (CT), imaging significantly depends on the attenuation coefficient. The attenuation coefficients are in turn dependent on the energy of the X-rays used to make the measurements. Imaging at a single energy produces an image of a single parameter. Due to this single energy spectrum, CT suffers from ambiguity and materials can appear identical. To eliminate this ambiguity and distinguish between the materials, dual-energy CT (DECT) is preferred. Moreover, dual-energy imaging is potentially more accurate than conventional single energy imaging for detecting lungs and bone-related diseases. In DECT, 20 to 40 keV is used for the lower energy band (i.e., imaging of soft tissue organs) and 120 to 140 keV is used for higher energy band (i.e., imaging of hard tissue organs). Therefore, the proposed CDIV-IVA (LoG) algorithm was tested on the dual-energy X-ray image in order to separate the ribs and lungs simultaneously. The dual-energy X-ray has the dual-energy PDF distribution. Hence, the proposed method is applied successfully and the obtained results are shown in Figure 2.13 for the separation of ribs and lungs.

The conventional chest X-ray has only single energy level and moreover only one image (mixture) is available for the BIS task. The proposed CDIV-IVA is also applied on the conventional X-ray to verify the ability to handle the underdetermined BSS case. However, for the IVA one input is a grayscale version of conventional X-ray and the other is a low contrast version of the first. Three conventional X-ray image sheets were collected from Kongunad Hospital and by using a standard camera picturized the films. The proposed algorithm successfully separated the lungs or

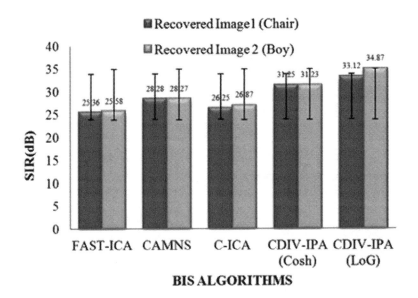

FIGURE 2.12
Comparison of SIRs in dB of demixed chair and boy images using different methods in non-negative BIS task.

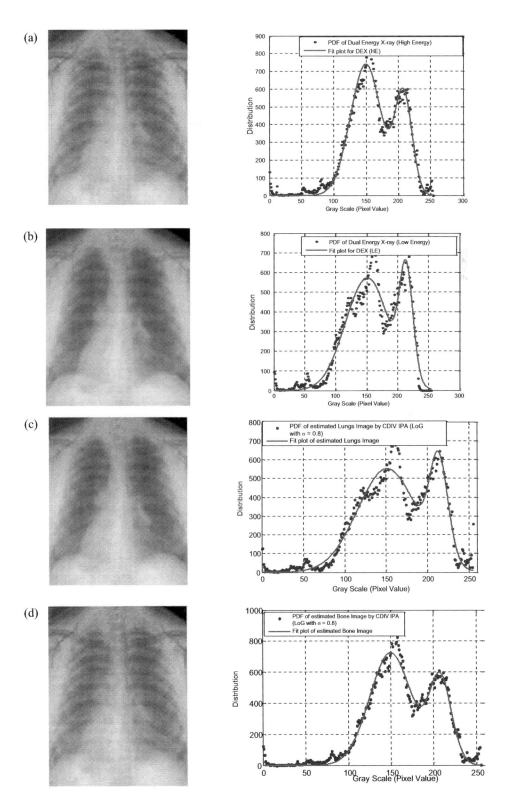

FIGURE 2.13
(a,b) The mixture of low and high-energy (dual-energy) chest X-ray images. (c–h) Separated lungs and bone images by CDIV-IVA (LoG) for (Sigma = 0.8, 1.4, and 1.8).

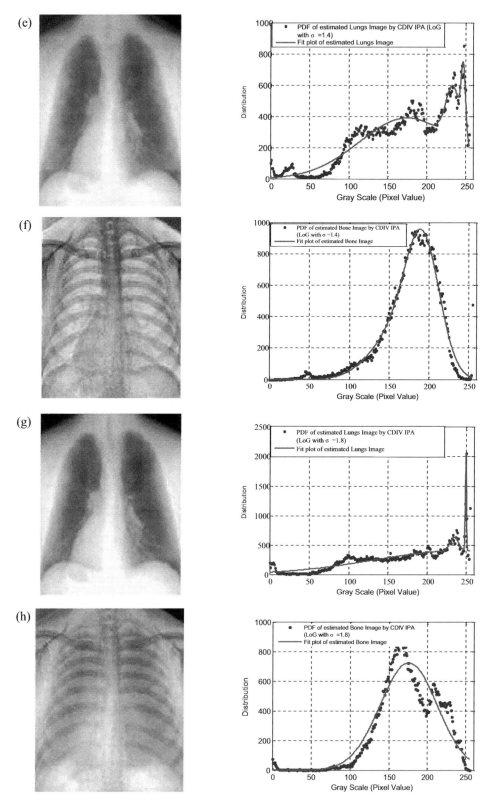

FIGURE 2.13
(Cont.)

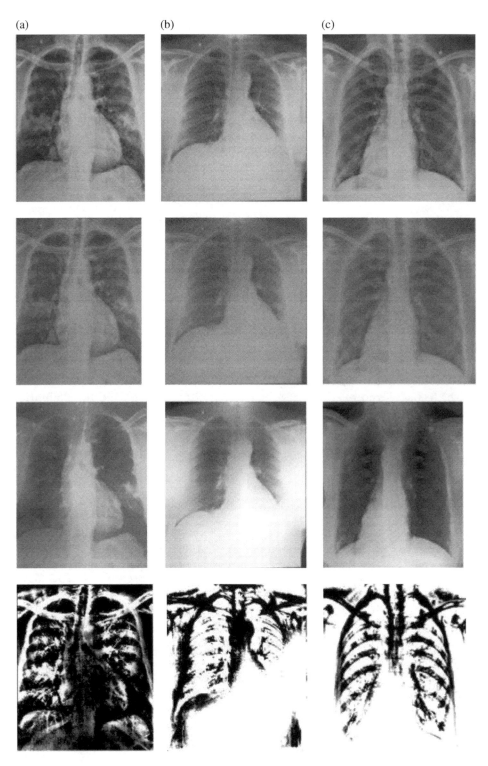

FIGURE 2.14
Separated lungs and bone images of conventional X-ray by CDIV-IVA (LoG). First row: conventional X-ray; second row: low contrast X-ray; third row: separated lungs images; and last row: separated bone images. a) Patient 1, b) Patient 2, and c) Patient 3.

soft tissues but not successful in the case of bony tissues. The obtained images are shown in Figure 2.14. Due to single energy spectrum of conventional X-ray, the separation of the bone tissue resulted in contrast shift or washout or binary imaging effect.

2.7 Conclusion

The following observations are obtained based on the experiments conducted on the non-negative image mixtures:

- Fast ICA has separated the source images with an overlapping effect. Moreover, the SIRs of the two estimated images were 5.89 dB and 155.65 dB lesser than CDIV-IVA (Cosh) and 7.76 dB and 9.29 dB lesser than CDIV-IVA (LoG).
- C-ICA has reasonably good (slightly better) separation performance. SIRs of the two estimated images were 5 dB and 4.36 dB lesser than CDIV-IVA (Cosh) and 6.87 dB and 8 dB lesser than CDIV-IVA (LoG).

CAMNS has perfectly separated the images, yet the SIRs of the two recovered images were 2.97 dB and 2.96 dB lesser than CDIV-IPA (Cosh) and 4.84 dB and 6.6 dB lesser than CDIV-IVA (LoG).

In separating the source images with non-negativity constraint, CDIV-IVA (LoG and Cosh) has achieved excellent results compared to the reported three algorithms.

References

Almeida M. & Almeida L., "Nonlinear separation of show-through image mixtures using a physical model trained with ICA," Signal Processing 92, no. 4 (2012): 872–884.

Anderson M., Xi-Lin L. & Adal. T., "Complex-valued independent vector analysis: Application to multivariate gaussian model," Signal Processing 92 (2012): 1821–1831.

Arai K., "Method for image source separation by means of independent component analysis: ICA, maximum entry method: MEM, and wavelet based method: WBM," International Journal of Advanced Computer Science and Applications 3, no. 11 (2012): 76–81.

Bronstein M., Zibulevsky M. & Zeevi Y., "Sparse ICA for blind separation of transmitted and reflected images," International Journal of Imaging Science and Technology 15, no. 1 (2005): 84–91.

Chan T.H., Ma W.K., Chi C.Y. & Wang Y., "A convex analysis framework for blind separation of non-negative sources," IEEE Transaction on Signal Processing 56, no. 10 (2008): 5120–5134.

Chen S., Suzuki K., Chan T. H. & Wang Y., "Separation of bones from chest radiographs by means of anatomically specific multiple massive-training ANNs combined with total variation minimization smoothing," IEEE Transactions on Medical Imaging 33 (2014): 246–257.

Chien J.T. & Hsieh H.L., "Convex divergence ICA for blind source separation," IEEE Transactions on Audio, Speech, and Language Processing 20, no. 1 (2012): 302–313.

Cichocki A. & Amari S., "Families of alpha- beta- and gamma- divergences: Flexible and robust measures of similarities," Entropy 12, no. 6 (2010): 1532–1568.

Comon P., "Independent component analysis, a new concept," Signal Processing 36, no. 3 (1994): 287–314.

Comon P. & Jutten C., Eds, Handbook of Blind Source Separation, Independent Component Analysis and Applications, Academic Press, Oxford, 2009.

Hwang H.S., Chung M.J., Kim S.M., Lee J. & Han H.,"A comparison between Dual-expose dual energy radiography and standard chest radiography for the diagnosis of small pulmonary nodules," Journal of the Korean Society of Radiology 59 (2008): 385–393.

Hyvarinen A., Karhunen J. & Oja E., Independent Component Analysis, John Wiley & Sons, New York, 2001.

Itahashi T. & Matsuoka K., "Stability of independent vector analysis," Signal Processing 92 (2012): 1809–1820.

Jenssen R., Principe J.C., Erdogmus D. & Eltoft T., "The Cauchy-Schwartz divergence and Parzen windowing: Connections to graph theory and Mercer kernels," Journal of the Franklin Institute 343 (2006): 614–629.

Kashani H., Gang G.J., Shkumat N.A., Varon C.A., Yorkston J.,Van Metter R., Paul N.S. & Siewerdsen J.H., "Development of a high-performance dual-energy chest imaging system, initial investigation of diagnostic performance," Academic Radiology 16 (2009): 464–476.

Lee I., Kim T. & Lee T.-W., "Fast fixed-point independent vector analysis algorithms for convolutive blind source separation," Signal Processing 87 (2007): 1859–1871.

Manji F., Wang J., Norman G., Wang Z. & Koff D., "Comparison of dual energy subtraction chest radiography and traditional chest X-rays in the detection of pulmonary nodules," Quantitative Imaging in Medicine and Surgery 6 (2016): 1–5.

Merrikh B. F., Babaie-Z. M. & Jutten C., "Linear-quadratic blind source separating structure for removing show-through in scanned documents," International Journal on Document Analysis and Recognition 14, no. 4 (2011): 319–333.

Noharaa T., Terao H., Tobe K., Musashi M. & Nagao K., "Risks of chest X-ray examination for students," Acta Medica Okayama 63 (2009): 43–47.

Oja E. & Plumbley M., "Blind separation of positive sources using non-negative PCA," Proceedings of the Fourth International Symposium on Independent Component Analysis, 2003: 11–15.

Rebuffel V. & Dinten J.M., (2007) "Dual-energy X-ray imaging: Benefits and limits," Insight -Non-Destructive Testing and Condition Monitoring 49, no 10 589-594.

Suzuki K., Abe H., Li F. & Doi K, "Suppression of the contrast of ribs in chest radiographs by means of massive training Artificial Neural Network," Proceedings of SPIE 5370 (2004): 1605–7422.

Tonazzini A., Salerno E. & Bedini L., "Fast correction of bleed through distortion in gray scale documents by a blind source separation technique," International Journal of Document Analysis and Recognition 10, no. 1 (2006): 17–25.

Wang F.Y., Chi C.Y., Chan T. H. & Wang Y., "Nonnegative least-correlated component analysis for separation of dependent sources by volume maximization," IEEE Transactions on Pattern Analysis and Machine Intelligence 32 (2010): 875–888.

Wang F.Y., Wang Y., Chan T.H. & Chi C.Y., "Blind separation of multichannel biomedical image patterns by non-negative least-correlated component analysis," Lecture Notes in Bioinformatics, Springer-Verlag 4146 (2006): 151–162.

Wang J. & Chang C.I., "Independent component analysis-based dimensionality reduction with applications in hyperspectral image analysis," IEEE Transaction on Geoscience and Remote Sensing 44, no. 6 (2006): 1586–1600.

Zhang H., Liping L. & Wanchun L., "Independent vector analysis for convolutive blind noncircular source separation," Signal Processing 92 (2012): 2275–2283.

Zymnis S.-J., Kim J.S., Parente M. & Boyd S., "Hyperspectral image unmixing via alternating projected subgradients," 41st Asilomar Conference on Signals, Systems, and Computers, Pacific Grove, 2007.

3

Rationalizing of Morphological Renal Parameters and eGFR for Chronic Kidney Disease Detection

Deepthy Mary Alex and D. Abraham Chandy

Electronics and Communication Department, Karunya Institute of Technology and Sciences, Karunya Nagar, Coimbatore, Tamil Nadu, India

Anand Paul

The School of Computer Science and Engineering, Kyungpook National University, Daegu, South Korea

3.1 Introduction

The present style of living in today's world has contributed to several types of diseases. Different types of diseases affect different organs in a human body in distinct number of ways. One such organ that is being affected in majority of the human race is the kidney. The kidneys are one of the vital and essential organs of the human system. Kidneys take a bean shape and are located on either side of the spine, below the ribs and behind the belly. The size of each kidney ranges from about 4 to 5 inches long and is approximately the size of a large fist. The main role of the kidney is to filter the blood. It functions to eliminate wastes, maintain the balance of body fluid, and control the levels of electrolytes. Several times a day the entire blood in the body passes through the kidneys. The fundamental process that occurs in the kidney is: Blood from different parts of the body comes into the kidney, where the unwanted elements are eliminated as waste and if necessary then salt, water and minerals are adjusted. The purified blood returns to the remaining parts of the body and the waste gets converted into urine and is discharged out of the body. Each kidney accounts for a million tiny filters named as nephrons. Nephrons are the momentous and crucial units of the kidney that function in filtration.

Each nephron consists of a renal corpuscle and a renal tubule. The renal corpuscle comprises of a cluster of capillaries known as glomerulus and an encompassing Bowman's capsule. From the capsule, a renal tubule enhances. Both the capsule and tubule are joined together and are comprised of epithelial cells with a lumen. Approximately about 0.8 to 1.5 million nephrons are present in each kidney of a normal healthy adult. Purification of the blood occurs as the blood passes through three layers, namely, the endothelial cells present in the capillary wall, basement membrane and between the foot processes of the podocytes of the lining of the capsule.

Chronic kidney disease (CKD) can be defined as a slow and progressive loss of kidney function which takes a period of several years. Ultimately, the patient will have permanent kidney failure. This kidney disease is irreversible or in other words it means the disease cannot be cured completely. The only possibility is to slow the progression of CKD to kidney failure by early detection and timely treatment. Due to high prevalence, morbidity, and mortality, CKD is considered as one among the important health problems of the public. According to studies (Abraham et al. 2016), when India is taken into consideration, India has less than 3% of land mass and anchors about 17% of the Earth's total population. Out of that, a huge number of patients below the poverty line, low gross domestic product and low monetary allocations for health care lead to minimal outcomes. Moreover, diseases like CKD have been often left ignored either because CKD is asymptomatic or because it cannot be cured completely. The burden on the poor patients to get timely treatment and routine check-ups will be high. Most often, CKD in the initial stages shows no or symptoms that are related to some other diseases, thereby leading CKD to advance stages. Some of the symptoms in the advanced stages are as follows: chest pain, dry skin, itching or numbness, feeling tired, headaches, increased or decreased urination, loss of appetite, muscle cramps, nausea, shortness of breath, sleep problems, trouble concentrating, vomiting, weight loss, etc. In most of the situation, the only way to check whether a person is suffering from CKD is by undergoing blood and urine tests that determine the kidney functioning. According to reports ("Chronic Kidney Disease (CKD) Surveillance Project" n.d.; "National Chronic Kidney Disease Fact Sheet 2017" n.d.), some of the people most prone to CKD are patients with:

- diabetes
- high blood pressure
- heart disease
- family history of kidney failure.

The basic test to detect CKD is by monitoring estimated glomerular filtration rate (eGFR) continuously over a period of three months. The normal range of eGFR for a healthy adult is from 100 to 130 mL/min/1.73 m^2. The standard measurement of eGFR is done in mL/min and is normalised to a body surface area of 1.73 m^2. If the eGFR is below the normal value for over three months, then it means that the kidney is not functioning normally and the patient has to undergo kidney ultra-sound scans to determine how far the kidney malfunctioning has affected the overall renal or kidney parameters. Table 3.1 shows the various stages of CKD based on eGFR.

Usually, eGFR is determined by taking blood samples. CKD can be detected using urine samples also. It is done by checking for abnormalities in albumin-to-creatinine ratio ("Know Your Kidney Numbers: Two Simple Tests" 2017) but it is not an accurate marker, and it is also tiring for patients since the patients have to collect urine hourly for 24 hours. Following are the main drive behind looking into the possibility of estimating eGFR using morphological renal parameters from ultrasound scans:

- Non-invasive
- Less cost, compared to carrying out blood test and then ultrasound
- Less stress to patients
- eGFR estimation time is less.

TABLE 3.1

Various Stages of Chronic Kidney Disease

Stage and description	eGFR value (ml/min/1.73 m^2)	Possible clinical symptoms
Stage 1: Kidney function normal with damage to kidney	90–120	Increased blood pressure, likely to be affected by urinary tract infections, unusual urinalysis
Stage 2: Mild kidney function loss with damage to kidney	60–89	Increased blood pressure, likely to be affected by urinary tract infections, unusual urinalysis
Stage 3a: Mild to moderate kidney function loss with damage to kidney	45–59	Increased blood pressure, likely to be affected by urinary tract infections, unusual urinalysis
Stage 3b: Moderate to severe kidney function loss with damage to kidney	30–44	Decrease in blood count values, loss of appetite, pain in the bones, unusual rise in nerve sensations, disturbed mental functioning, feeling dizzy
Stage 4: Severe kidney function loss with severe damage to kidney	15–29	Swelling in legs, decrease in blood count values, loss of appetite, unusual rise in nerve sensations, disturbed mental functioning, feeling dizzy
Stage 5: Complete damage to kidney leading to kidney failure known as end stage renal disease (ESRD)	Less than 15	Shortness of breath, loss of appetite, tiredness, mentally and physically weak

3.2 Typical GFR Estimation Methods Used

Currently, there are various methods to estimate eGFR. eGFR is calculated by taking the blood samples of patients and estimated using equations. The equations to estimate eGFR are presented in this chapter. Measurement of GFR cannot be taken directly since filtration occurs parallel in millions of glomeruli present in both the kidneys. However, by calculating the rate at which a marker injected into the bloodstream is eliminated as urine can determine GFR. Inulin is considered as one such marker. Inulin is a polysaccharide that can neither be secreted nor reabsorbed by the kidney instead filtered through the renal glomeruli, thereby making inulin a gold standard to measure GFR. The main drawback of this technique is that it is complicated and expensive. Other substances that are used to measure GFR include renal or plasma clearances of diethylene-triamine penta-acetic acid (DTPA), radioactively tagged ethylenediaminetetra-acetic acid (51Cr-EDTA), iohexol (X-ray contrast medium), iothalamate (radioactively tagged contrast medium), plasma clearance of inulin, and endogenous clearance of creatinine (renal clearance of creatinine). A definite view as to which substance among the above is gold standard in measuring GFR is yet to be identified.

Another alternative to the problem stated is to estimate GFR instead of measuring GFR. It has been found that the most common method for estimating GFR is based on creatinine. It is seen that when GFR decreases, the plasma concentration of creatinine increases thereby giving a rough indicator that the kidneys are impaired. Even though this method is often used, it has its own drawbacks too. It is proved that the plasma concentration of creatinine depends on the muscle mass, age, sex, weight, height, etc.

Hence, it is an open question as to which creatinine equation yields an accurate estimate of GFR in different situations and different orders.

Apart from using creatinine, cystatin C is used as a marker for estimating GFR. Cystatin C can be defined as a low-molecular-weight protein produced in all the cells of the human body having a nucleus. As stated before, glomeruli also filters cystatin C freely without secretion or reabsorption. The advantage of cystatin C over creatinine is that it doesn't depend on parameters such as age, sex, weight, height, etc. Thus, cystatin C is also being used widely in many places. It is still a controversy as to when and in which situation cystatin C should be used to estimate GFR.

Apart from the equations based on the stated markers, equations based on the combination of both the markers have also been developed to estimate GFR. This equation can either represent the mean of estimations of GFR with either marker or the composite equation containing both the markers.

Given next are the different equations to estimate GFR based on creatinine, cystatin C, and combination of both (Florkowski and Chew-Harris 2011; Inker et al. 2012; Stevens et al. 2008; Zaman, n.d.). From Equations (3.1) to (3.7), eGFR is determined based on serum creatinine. Equation (3.8) is used to estimate GFR using cystatin C and finally Equation (3.9) uses both serum creatinine and cystatin C to determine GFR.

i. Equations based on serum creatinine:

1. Modification of diet in renal disease (MDRD)

$$eGFR = 186 \times (S.Cr)^{\frac{1}{1.154}} \times \left(Age\right)^{\frac{1}{0.203}} \times (0.742\,\text{for female}) \times (1.210\,\text{for black}) \qquad (3.1)$$

where S. Cr is serum creatinine measured in mg/dL.

2. Chronic kidney disease – epidemiology collaboration (CKD-EPI) creatinine equation

a) For females with S. Cr ≤ 62 µmol/L

$$eGFR = (144 + 22\,\text{if black}) \times \left(\frac{Cr}{0.7}\right)^{\frac{1}{0.329}} \times (0.993)^{Age} \qquad (3.2)$$

b) For females with S. Cr > 62 µmol/L

$$eGFR = (144 + 22\,\text{if black}) \times \left(\frac{Cr}{0.7}\right)^{\frac{1}{1.209}} \times (0.993)^{Age} \qquad (3.3)$$

c) For males with S. Cr ≤ 80 µmol/L

$$eGFR = (141 + 22\,\text{if black}) \times \left(\frac{Cr}{0.9}\right)^{\frac{1}{0.411}} \times (0.993)^{Age} \qquad (3.4)$$

d) For males with S. Cr > 80 μmol/L

$$eGFR = (141 + 22 \text{ if black}) \times \left(\frac{Cr}{0.9}\right)^{\frac{1}{1.209}} \times (0.993)^{Age} \tag{3.5}$$

where S. Cr is serum creatinine measured in μmol/L.

3. Cockcroft–Gault equation

a. For females

$$eGFR = (140 - Age) \times Weight \times \left(\frac{1.04}{S.Cr}\right) \tag{3.6}$$

b. For males

$$eGFR = (140 - Age) \times Weight \times \left(\frac{1.23}{S.Cr}\right) \tag{3.7}$$

where S. Cr is measured μmol/L and weight in kilograms.

ii. Equations based on cystatin C:
   1. CKD-EPI cystatin C equation:

$$eGFR = 133 \times Min\left(\frac{S_{cys}}{0.8 \text{ or } 1}\right)^{\frac{1}{0.499}} \times Max\left(\frac{S_{cys}}{0.8 \text{ or } 1}\right)^{\frac{1}{1.328}}$$
$$\times (0.996)^{Age} \times (0.932 \text{ if female}) \tag{3.8}$$

where S_{cys} (standardized serum cystatin C) is measured in mg/L, Min indicates the minimum of $S_{cys}/0.8$ or 1, Max represents maximum of $S_{cys}/0.8$ or 1, and age is given in years.

iii. Equations based on both serum creatinine and cystatin C:
   1. CKD-EPI creatinine-cystatin equation

$$eGFR = 135 \times Min\left(\frac{S_{cr}}{K,1}\right)^{\alpha} \times Max\left(\frac{S_{cr}}{K,1}\right)^{\frac{1}{0.601}} \times Min\left(\frac{S_{cys}}{0.8,1}\right)^{\frac{1}{0.375}}$$
$$\times Max\left(\frac{S_{cys}}{0.8,1}\right)^{\frac{1}{0.711}} \times (0.955)^{Age} \times (0.969 \text{ if female}) \times (1.08 \text{ if black}) \tag{3.9}$$

where S_{Cr} (serum creatinine) and S_{cys} (standardized serum cystatin C) are measured in mg/L, K is taken as 0.7 for females and 0.9 for males, $Min\left(\frac{S_{cr}}{K,1}\right)$ indicates the minimum of S_{Cr}/K or 1, $Max\left(\frac{S_{cr}}{K,1}\right)$ indicates the maximum of S_{Cr}/K or 1, $Min\left(\frac{S_{cys}}{0.8,1}\right)$ indicates the minimum of $S_{cys}/0.8$ or 1, Max indicates maximum of $S_{cys}/0.8$ or 1, age is given in years, and α is taken as −0.248 for females and −0.207 for males.

From the above equations, it is seen that estimated GFR is based on conventional or clinical parameters of kidney, that is, the equations depend on the blood samples. In the next section, the possibility of estimating GFR using nonconventional or morphological parameters of kidney based on image processing of kidney ultrasound images is considered.

3.3 Kidney Morphology Toward eGFR

Morphological renal parameters refer to the measurements of kidney that help in discriminating between chronic and acute failure or parameters that define the functionality of the kidney, that is, how well the kidney is functioning. Chronic failures are not reversible whereas acute failure is reversible. The morphological parameters of the kidney are:

- Renal length
- Total kidney volume
- Cortical thickness
- Parenchymal thickness
- Echogenicity
- Corticomedullary differentiation.

The kidney has an oval bean shape and it is mostly viewed in the longitudinal plane scan. According to studies (Hansen, Nielsen, and Ewertsen 2015; O'Neill 2014), it is seen that right kidney is caudal and slimmer compared to left kidney which may be due to the so-called dromedary hump due to its proximity to the spleen. A capsule surrounds the kidney thereby separating it from perirenal fat. The kidney can further partition into parenchyma and renal sinus. It is found that renal sinus is hyperechoic and comprises of calyces, the renal pelvis, fat, and the major intrarenal vessels. In a normal kidney, the urinary collecting system present in the renal sinus is unseen, but it develops a heteroechoic appearance with interposed fat and vessels. Compared to renal sinus, the parenchyma is more hypoechoic and homogenous. Parenchyma is further divided into the outermost cortex and the innermost and slightly less echogenic medullary pyramids.

The renal length of a normal adult ranges from 10 to 12 cm, and it is seen that left kidney is often slightly shorter than the right kidney. The cortical thickness ranges from 7 to 10 mm and is estimated from the base of the pyramid. If in case the pyramids are difficult to differentiate, then parenchymal thickness can be measured. The range of parenchymal thickness should be 15–20 mm. In Figure 3.1, the measurements of the kidney are represented. According to studies and works carried out, it is seen that echogenicity of the cortex decreases with age.

Echogenicity can be defined as the amount of sound that is being reflected back to the probe; it is dependent on the amplitude of incident sound, how much of the sound is absorbed, how much is reflected and the angle of reflection. Evaluation of only cortical echogenicity can be done and qualitatively it should be less than the liver or spleen. Corticomedullary differentiation is the ability to differentiate the cortex and medullae but in some cases it is difficult to view the medullae, hence difficult to differentiate from

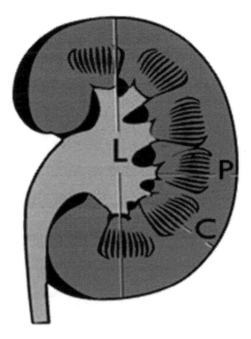

FIGURE 3.1
Measurements of kidney: L-Renal length, C-Cortical thickness, P-Parenchymal thickness (Source: Hansen et al., 2015- with permission).

ultrasound images even in healthy adults. It is to be noted that the prominence of the medullae is usually abnormal, generally indicating increased echogenicity of the cortex. Total kidney volume is calculated using the ellipsoidal formula. In the longitudinal scan of the kidney, the two poles, that is, the superior and inferior poles are identified. The renal length (L) is determined as the longest distance between the two poles. The thickness or the anteroposterior diameter (AP) is also determined in the longitudinal scan. AP is marked as the maximum distance between the anterior and posterior walls at the mid-third of the kidney. For calculating the renal width (W), transverse scan is considered. W is the maximum transverse diameter and it was taken at the hilium. Centimeter (cm) is taken as the unit of measurement. Total volume can be determined using the following equation (Kim et al. 2008):

$$\text{Kidney volume} = L \times AP \times W \times 0.523 \qquad (3.10)$$

3.4 Endorsing Factors toward Image-Based GFR Estimation

All the morphological parameters stated are determined using different imaging modalities such as magnetic resonance imaging (MRI), ultrasound (US), and computed tomography (CT) on the abdomen of a patient. Here, US is only considered because it is preferred as the safest modality for scanning since the ultrasound waves are harmless compared to CT and MRI. Other advantages of US over the other two modalities are: It offers real-time, cheaper, safest, and least invasive imaging technology for doing diagnosis and therapeutic procedures. In India, doctors prescribe US scanning for CKD detection because most of the patients are below poverty line and other two modality examination is very expensive. There are quite a number of literatures that propose one or more morphological parameters correlate with serum creatinine or eGFR, but till now no formula relating eGFR and morphological parameters determined from an US scan of a kidney has been developed.

Yoruk, Hargreaves, and Vasanawala (2018) propose an equation for eGFR based on kidney volume and transfer coefficient (K_{trans}) from MRI, but Egberongbe et al. (2010) explain that the kidney volume depends on anthropometric parameters such as age, sex, height, weight, body surface area (BSA), body mass index (BMI), etc. Hansen, Nielsen, and Ewertsen (2015) emphasize that the adult kidney size a is variable in correlation with body height and age. Singh et al. (2016) proposed that there is a significant correlation between echogenicity and serum creatinine. Sanusi et al. (2009) stated that kidney volume correlate with eGFR. It is evident from the study of Beland et al. (2010) that cortical thickness also correlates with eGFR and finally, Shivashankara (2016) proposes that there is a linear correlation between parenchymal thickness and eGFR.

The morphological renal parameters can either be determined manually, semiautomatically, or automatically from an ultrasound image. Once these parameters are determined, then it will be easy to detect and grade CKD based on the abnormal values of morphological parameters and eGFR from the image.

3.5 Conclusion

From the previous sections, it is clear that the possibility of estimating GFR from an US image is high. The equation should be dependent on the morphological renal parameters and anthropometric parameters keeping in mind that these parameters vary from region to region, that is, the body stature of an Indian male will not be the same as that of an African male.

As already stated, Yoruk, Hargreaves, and Vasanawala (2018 had proposed an eGFR equation based on only the kidney volume and transfer coefficient but patients with CKD were not considered to estimate GFR. Hence, the eGFR determined for CKD patients using the proposed equation is still a doubt. Another concern is, the author proposed the equation for dynamic contrast enhanced (DCE)-MRI. Whether it is possible to apply the proposed equation on ultrasound segmented kidneys is still a matter in question. Detection of CKD using MRI scan is quite expensive, in fact expensive than CT scan. In western countries, the citizens are health insured by the government, hence cost of imaging is not a matter of concern. In developing countries like India, the patient has to pay for their own health treatments, hence Indian doctors suggest only ultrasound scan for CKD detection. MRI scan provides better precision but cost is prohibitive and is not freely available. CKD alters one or more morphological parameters, hence while developing eGFR equation, the morphological parameters that correlate with eGFR from the existing literatures should be considered. Some of the alterations occurring on the kidney of CKD patients are as follows: (i) Renal length and kidney volume reduces, (ii) echogenicity increases, (iii) cortical thickness reduces, and (iv) as CKD progresses difficulty in corticomedullary differentiation increases. The morphological parameters of kidney also depend on anthropometric parameters, that is, a very short normal person will have renal length less than normal. Depending upon the age of a person also changes occur in the kidney, that is, an old person might have kidney volume less than that of the normal range. To estimate GFR from ultrasound kidney images, anthropometric parameters are relevant.

It is a necessity to estimate GFR from ultrasound kidney scans for the detection of CKD. The possibility of a novel and cost-effective method will be great relief for the patients suffering from CKD in terms of stress when taking blood sample and urine sample, money, and time. Finally, to conclude, the study on morphological renal parameters and eGFR is reproducible and will provide a platform for a cheaper source of GFR estimation in CKD patients on routine basis. However, to come up with an appropriate equation, the need for a large sample consisting of patients and nonpatients is important.

References

Abraham, Georgi, Santosh Varughese, Thiagarajan Thandavan, et al. 2016. "Chronic Kidney Disease Hotspots in Developing Countries in South Asia." *Clinical Kidney Journal* 9 (1): 135–141. doi:10.1093/ckj/sfv109.

Beland, Michael D., Nicholas L. Walle, Jason T. Machan, and John J. Cronan. 2010. "Renal Cortical Thickness Measured at Ultrasound: Is It Better than Renal Length as an Indicator of Renal Function in Chronic Kidney Disease?" *AJR. American Journal of Roentgenology* 195 (2): W146–W149. doi:10.2214/AJR.09.4104.

"Chronic Kidney Disease (CKD) Surveillance Project." n.d. Accessed September 16, 2018. https://nccd.cdc.gov/ckd/default.aspx.

Egberongbe, Adedeji A., Victor A. Adetiloye, Abiodun O. Adeyinka, Olusegun T. Afolabi, Anthony O. Akintomide, and Olugbenga O. Ayoola 2010. "Evaluation of Renal Volume by Ultrasonography in Patients with Essential Hypertension in Ile-Ife, South Western Nigeria." *Libyan Journal of Medicine* 5 (1): 4848. doi:10.3402/ljm.v5i0.4848.

Florkowski, Christopher M, and Janice SC Chew-Harris. 2011. "Methods of Estimating GFR – Different Equations Including CKD-EPI." *The Clinical Biochemist Reviews* 32 (2): 75–79.

Hansen, Kristoffer, Michael Nielsen, and Caroline Ewertsen. 2015. "Ultrasonography of the Kidney: A Pictorial Review." *Diagnostics* 6 (1): 2. doi:10.3390/diagnostics6010002.

Inker, Lesley A., Christopher H. Schmid, Hocine Tighiouart, et al. 2012. "Estimating Glomerular Filtration Rate from Serum Creatinine and Cystatin C." *New England Journal of Medicine* 367 (1): 20–29. doi:10.1056/NEJMoa1114248.

Kim, Hyun Cheol, Dal Mo Yang, Sang Ho Lee, and Yong Duck Cho. 2008. "Usefulness of Renal Volume Measurements Obtained by a 3-dimensional Sonographic Transducer with Matrix Electronic Arrays." *Journal of Ultrasound in Medicine* 27 (12): 1673–1681. doi:10.7863/jum.2008.27.12.1673.

"Know Your Kidney Numbers: Two Simple Tests." 2017. National Kidney Foundation. January 30, 2017. www.kidney.org/atoz/content/know-your-kidney-numbers-two-simple-tests.

"National Chronic Kidney Disease Fact Sheet, 2017." n.d., 4.

O'Neill, W. Charles. 2014. "Renal Relevant Radiology: Use of Ultrasound in Kidney Disease and Nephrology Procedures." *Clinical Journal of the American Society of Nephrology* 9 (2): 373–381. doi:10.2215/CJN.03170313.

Sanusi, Abubakr A., Fatiu A. Arogundade, Olusola C. Famurewa, et al. 2009. "Relationship of Ultrasonographically Determined Kidney Volume with Measured GFR, Calculated Creatinine Clearance and Other Parameters in Chronic Kidney Disease (CKD)." *Nephrology, Dialysis, Transplantation: Official Publication of the European Dialysis and Transplant Association – European Renal Association* 24 (5): 1690–1694. doi:10.1093/ndt/gfp055.

Shivashankara, Vinayaka Undemane. 2016. "A Comparative Study of Sonographic Grading of Renal Parenchymal Changes and Estimated Glomerular Filtration Rate (EGFR) Using Modified Diet in Renal Disease Formula." *Journal of Clinical and Diagnostic Research* doi:10.7860/JCDR/2016/16986.7233.

Singh, Arvinder, Kamlesh Gupta, Ramesh Chander, and Mayur Vira. 2016. "Sonographic Grading of Renal Cortical Echogenicity and Raised Serum Creatinine in Patients with Chronic Kidney Disease." *Journal of Evolution of Medical and Dental Sciences* 5 (May). doi:10.14260/jemds/2016/530.

Stevens, Lesley A., Josef Coresh, Christopher H. Schmid, et al. 2008. "Estimating GFR Using Serum Cystatin C Alone and in Combination with Serum Creatinine: A Pooled Analysis of 3,418 Individuals with CKD." *American Journal of Kidney Diseases: the Official Journal of the National Kidney Foundation* 51 (3): 395–406. doi:10.1053/j.ajkd.2007.11.018.

Yoruk, Umit, Brian A. Hargreaves, and Shreyas S. Vasanawala. 2018. "Automatic Renal Segmentation for MR Urography Using 3d-GrabCut and Random Forests." *Magnetic Resonance in Medicine* 79 (3): 1696–1707. doi:10.1002/mrm.26806.

Zaman, Sojib Bin. n.d. "Detection of Chronic Kidney Disease by Using Different Equations of Glomerular Filtration Rate in Patients with Type 2 Diabetes Mellitus: A Cross-Sectional Analysis." *Cureus* 9 (6). Accessed September 16, 2018. doi:10.7759/cureus.1352.

4

Human Computer Interface for Neurodegenerative Patients Using Machine Learning Algorithms

S. Ramkumar

School of Computing, Kalasalingam Academy of Research and Education, Krishnankoil, Virudhunagar (Dt), India

G. Emayavaramban

Department of Electric and Electronic Engineering, Karpagam Academy of Higher Education, Coimbatore, India

J. Macklin Abraham Navamani

Department of Computer Applications, Karunya Institute of Technology and Sciences, Coimbatore, India

R. Renuga Devi, A. Prema, and B. Booba

School of Computing Science, Vels Institute of Science,Technology and Advanced Studies (VISTAS), Pallavaram, Chennai, India

P. Sriramakrishnan

School of Computing, Kalasalingam Academy of Research and Education, Krishnankoil, Virudhunagar (Dt), India

4.1 Introduction

A neurodegenerative disease causes disabilities for normal persons. Persons with disabilities are unable to share their ideas as well as thoughts with others; they need some assistance from known person with the help of new devices with trendy technologies for their communication and mobility. An individual person with disability condition is called locked-in state. To stay away from this state, several researchers tried to find a solution for disabled persons to develop a smart system in terms of communication and action to enhance the value of their lifestyle without assistance [1]. Currently, we live in a technological world. The technology that we live with today is powerful, fast, and sophisticated. By using technology we are able to communicate with others through interfaces like human computer interface or brain computer interfaces in the absence of biological channel with the help of bio signals. According to WHO, there are different types of neuromuscular disorders which kills the neurons and make a person unable to communicate or walk; for such persons eye activities play a vital role to connect to the external devices to share their thoughts. So the researchers turned their attention toward alternative communication system for the elderly disabled to overcome their problem in a natural way. The alternative communication mediums are categorized into hand signals, mental thoughts, and eye movement activities. Hand signals are possible only for the individual who can talk but think about the individual who have no speech capability due to quadriplegic cerebral palsy, while brain activity or event-related potentials require more practice and method is slow. Eye movement function has received attention in the search of nonconventional ways of controlling computer techniques using EOG. HCI is the study of the translating human action into definite pattern. These patterns are converted into commands [2–4]. This study concentrates on identifying and recognizing eye movement techniques as more suitable for designing multi states Human computer interaction systems using 11 different eye movements.

The inspiration behind the study is to create nine states HCI based on eye activities. Currently, several researchers focused on developing variety of applications using EOG. Some of them are health monitoring system [5], cursor control system [6], controlling home appliances [7], wheelchair [8], hospital alarm system [9], lip movement control system [10], sip-and-puff controller [11], virtual keyboard [12], television control system [13], and controlling medical instrument [14]. The surveys collected from different research papers have used four states to develop the HCI using eye activities. This chapter determines the best method for designing a nine states HCI using eight (events) and three (nonevents) eye movements.

This chapter is categorized into six sections. Section 4.1 introduces needs of EOG-based HCI while Section 4.2 presents in details about literature surveys in the domain of EOG-based HCI. In Section 4.3, we sum up the methodology used in this research and Section 4.4 explains the signal classification techniques. Analysis of result for both network models are discussed in Section 4.5 and Section 4.6 is the Conclusion.

4.2 Background

EOG signals collected from human subjects are used in biomedical engineering for developing assistance for disabled persons. Different types of well-organized assistive devices are developed in the past. Some of the outstanding researches are explained in the study. Sun et al., discussed bio-based HCI for developing assistive robot using EOG signals collected from four volunteers (3M, 1 F) using five electrodes is discussed. Entropy, spatial entropy, and peak-to-valley ratio are used to compose feature vectors that have less feature dimensions. Second, the back-propagation (BP) neural network, classification support vector machine (C-SVM), and particle swarm optimization-SVM (PSO-SVM) have been used to classify four kinds of eye movement signals like right, left, blink once, and blink twice. The back propagation and C-SVM classifiers yield the same classification accuracy rate of approximately 96%, but the hybrid PSO-SVM obtains the highest classification accuracy rate of 98.71% and exhibits more stable classification performance than the BP network [15]. Wissel.T. and Palaniappan.R, designed HCI based on electrooculogram signal acquired from five subjects using five-electrode system for developing virtual keyboard. Acquired signals are applied to wavelet decomposition for extracting features. Linear Discriminant Analysis (LDA) method is used to classify the EOG signals. Experiment was conducted for six tasks, namely, right, left, up, down, long blink and short blink with an average accuracy of 94.90% [16]. Postelnicu.et al. present an EOG-based interface for HCI to control keyboard, mouse, and joystick. Signals are collected from 14 subjects aged from 23 to 30 using five-electrode systems for four tasks to extract features using peak amplitude algorithm for both horizontal and vertical eye movements. Obtained features are classified with fuzzy logic and deterministic finite automaton (DFA) to obtain an average accuracy of 95.63% with sensitivity and specificity of 97.31% and 93.65%[17]. R.Barea et al. describe a new eye-control method for eye-based computer interaction using EOG using wavelet transform and radial basis function. Signals were collected from ten human subjects. The results obtained demonstrate that system's reliability in detecting eye movements is 100% for saccadic movements [18].A.Banerjee designed and developed a wheelchair using EOG signals collected from five subjects aged between 23 and 25 using three-electrode systems for five tasks. Acquired signals are applied to the discrete wavelet transform to extract prominent features. Extracted features are compared through several classifiers, namely, LDA, QDA, SVM, FFNN, KNN, and RBF. After comparison the experiment result shows that RBF outperforms remaining classifier with average mean accuracy of 77.36% [19]. Nakanishi et al. identify voluntary eye blink detection method using electrooculogram for controlling brain computer interface using normal blink, double blink, and wink from vertical and horizontal EOG signals collected from eight subjects using five-electrode system. Acquired EOG signals were applied to vertical peak amplitude; horizontal peak amplitude and maximum cross correlation extract the features. Eye blinks were classified by support vector machine. As the result of simulations, an average accuracy of 97.28% was obtained using this method [20]. Nathan et al. designed and developed an EOG-based typing system for typing letters on the monitor using eight different tasks by extracting the features with the help of amplitude and timing of positive and negative components of the signal. From this study the designed system shows an average typing speed of 15 letters per minute with a classification accuracy of 95.2% compared with the existing method's classification accuracy of 90.4% [21]. Sherenby et al. introduced a new method for designing HCI to control the mouse and the document reader for handicapped persons to control the applications using EOG signals collected from five subjects. Numerical integration techniques are applied to extract the feature, and neural network is used to classify the data obtained from the different subjects and attain 90% of accuracy [22]. The majority of the study on EOG-based HCI spotlight on limited states varying from two to five, but this study explores the possibility of increasing the number of states of HCI to nine states using controllable eye movements.

4.3 Protocol and Signal Acquisition

Sixteen eye movements are studied in the beginning and it is concluded that eleven eye movements could be voluntarily controlled, the remaining five eye movements

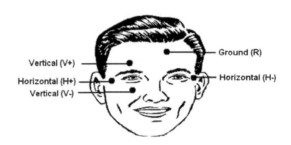

FIGURE 4.1
Electrode placement for EOG signal acquisition.

are not easy to be controlled by all subjects. From the chosen eleven movements eight movements are considered as events and three are taken as nonevents to design nine states HCI. Protocol designs for signal acquisition and noise removal are explained by the same author in the previous study [23]. Electrode placements and spectral analysis for subject S8 are depicted in Figures 4.1 and 4.2, respectively.

The unprocessed signals from the subjects are applied with short time Fourier transform to observe the frequency level of each task per ten trials. From this analysis, we found that nonevents task frequency is high compared to event tasks performed by individuals and fall in between 6Hz and 9Hz while event frequency level is between 5Hz and 9Hz, which is shown in Table.4.1.

4.3.1 Feature Extraction

After the spectrum identification, collected signals are applied to Parseval theorem to collect the prominent features. The feature extraction algorithm consists of the following steps:

Step 1: S = Sample data of two channel EOG signal for 2 s.

Step 2: S is partitioned into 0.1 s windows.

Step 3: Bandpass filters are applied to extract eight frequency bands from S.

Step 4: Apply Fourier transform to the frequency band signal to extract the features.

Step 5: Extract the absolute values and sum of the power values are extracted.

Step 6: Take the average values from each frequency band.

Using this concept total energy of the EOG signal is extracted for all the twenty subjects. 110 data samples for one subject were obtained. The feature sets obtained from the above feature extraction method are individually applied to neural networks as input features to classify the signals into eleven different eye movement tasks. The feature sets obtained from single trial for each task for one subject is demonstrated in Figure 4.3.

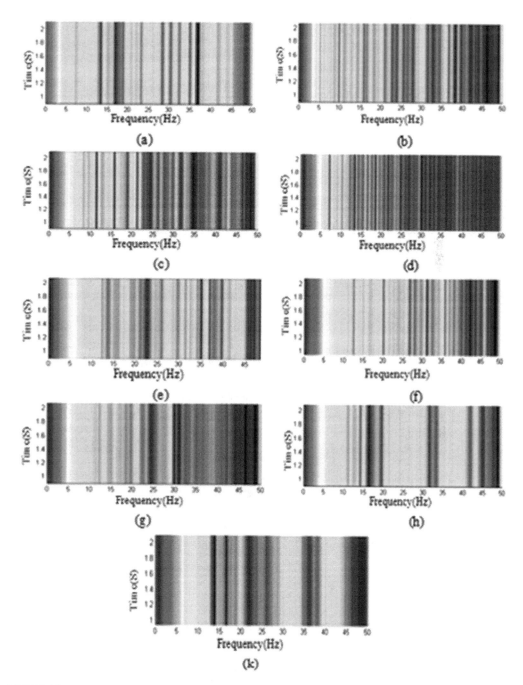

FIGURE 4.2

Spectral analysis for different eye movements: (a) right, (b) left, (c) up right, (d) down right, (e) up left, (f) down left, (g) rapid movement, (h) lateral movement, (i) open, (j) close, (k) stare of subject S8.

TABLE 4.1

Average frequency of twenty subjects

Task	Average Frequency (Hz)
Right	5–6
Left	5–6
Up Right	7–8
Down Right	6
Up Left	7–8
Down Left	7
Rapid Movement	8
Lateral Movement	6
Open	9
Close	6–8
Stare	8–9

4.4 Signal Classification

To classify the EOG signals collected from different tasks, two dynamic neural network models are designed. This study uses the layered recurrent network and distributed time delay network to categorize the EOG data signals. Performances and recognition accuracy for the two dynamic networks are compared using Parseval features to validate the result. The networks are designed with sixteen input neurons and four output neurons. Each network is tested with 110 data samples gathered from single subjects for the eleven eye movement task. Neurons in the hidden layer are chosen experimentally and fixed at eight. Both networks were trained using Levenberg back-propagation training algorithm. Experiment study proves that the performance of the network is better with eight hidden neurons. All the samples are normalized between 0 to 1 using a binary normalization algorithm to fit the data within range of 1. Out of the 110 samples, 75% of the data are used in the training of the network [23]. Sixteen input features and eight hidden neurons with four output neurons are used to classify the events and nonevents performed by different subjects. The learning, maximum iteration and testing error tolerance were already discussed by the same authors [23–27].

4.4.1 Layered Recurrent Neural Network

A layered recurrent neural network (LRNN) is a dynamic network, which produces an internal state of the network that permits to demonstrate dynamic maps between input and output. When the output of a neuron is feedback into a neuron in earlier layer, the output of that neuron is a function of both inputs from previous layer that existed for one cycle of calculation. Each input units are fully associated to hidden units and output units. Unlike feed forward neural networks, LRNN can use their internal memory to process arbitrary sequences of inputs [24–27]. Figure 4.4 shows the structure of the LRNN applied in this experiment.

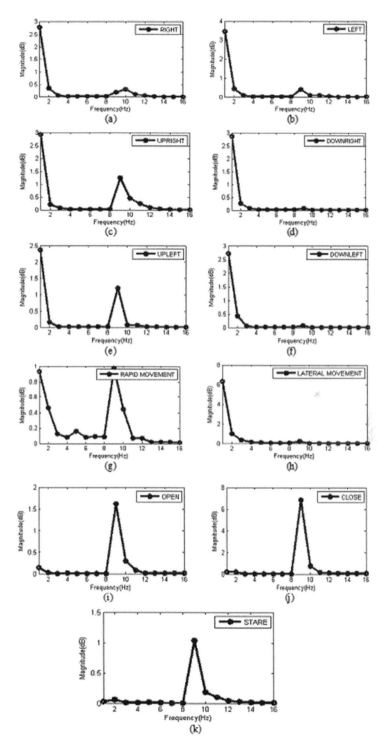

FIGURE 4.3
Feature extracted signal from eleven different eye movements:(a) right, (b) left, (c) up right, (d) down right, (e) up left, (f) down left, (g) rapid movement, (h) lateral movement, (i) open, (j) close, (k) stare of subject S8 using Parseval theorem.

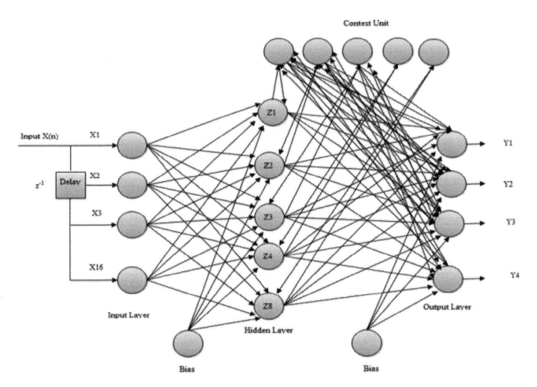

FIGURE 4.4
Layered recurrent neural network model.

4.4.2 Distributed Time Delay Neural Network

A distributed time delay neural network (DTDNN) is one of the effective dynamic networks. Compared to static network it is more effective because it has dynamic memory space to study the time varying and sequential patterns. The working principle of DTDNN is similar to feed forward network except that each input and layer weights have a tap delay line associated with each input. Therefore, the network has a tapped delay line that senses all the signals before it is connected to the network weight matrix through delay time units in ascending order from left to right correspond to the weight matrix [24–35]. Figure 4.5 shows the architecture of the DTDNN model used in this study.

4.5 Result

We conducted our experiment with ADI T26 Power Lab by using two-channel arrangements to measure horizontal and vertical eye movements using cup-shaped electrodes placed on the human face. Twenty subjects participated in the experiment. Through our experiment we compared two dynamic neural networks using Parseval theorem. Classification accuracy of the dynamic network models for the eleven eye movement is shown

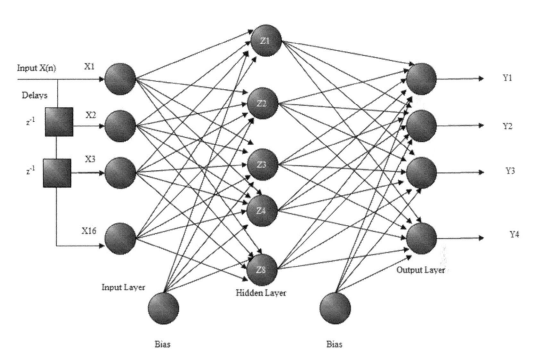

FIGURE 4.5
Distributed time delay neural network model.

in Table 4.2. Table 4.2 displays the accuracy of the LRNN model and DTDNN model for Parseval features correspondingly. From Table 4.2, it is observed that the Parseval features using LRNN have been trained with a mean training time of 34.67 seconds and mean testing time of 0.78 seconds. The performance of the classification system has the mean maximum classification accuracy of 94.64% and the mean minimum classification accuracy of 87.43%. The maximum mean classification accuracy rate is observed for S14 at 92.50% is achieved. The overall classification accuracy of 91.13% has been obtained, while DTDNN has been trained with a mean training time of 3.40 seconds and mean testing time of 0.61 seconds. The performance of the classification system has the highest mean classification accuracy of 94.37% and the average minimum classification accuracy of 86.80%.The maximum mean classification rate is observed for S14 at 91.44% is achieved. The overall classification accuracy of 90.81% has been obtained. Analyzing the performance of forty network models it is seen that performance of the Parseval features using LRNN outperforms the DTDNN in terms of minimum average, maximum average, and average mean classification accuracy. The classification accuracy for Parseval features using LRNN and DTDNN is shown in Figure 4.6. The experimental result proved that Parseval features using LRNN model are more appropriate for identifying the EOG signals for eleven eye movement tasks to model the HCI.

From Table 4.3 it is concluded that maximum average recognition accuracy of above 85% is achieved by the subject S8, second maximum average recognition accuracy of 81% is achieved by the subject S9,minimum average recognition accuracy of 72% is achieved by subject S4, and the remaining subjects average recognition accuracy of 74% is achieved using Parseval features with LRNN architectures as shown in Figure 4.7.

TABLE 4.2

Classification performance of LRNN and DTDNN using Parseval features

Sub	Mean Training Time (s)	Mean Testing Time (s)	Classification Performance for LRNN (%)				Mean Training Time (s)	Mean Testing Time (s)	Classification Performance for DTDNN (%)			
			Max	Min	Mean	SD			Max	Min	Mean	SD
S1	32.84	0.62	94.55	88.18	90.94	1.66	3.62	0.65	93.64	87.27	90.70	1.98
S2	32.81	0.64	94.55	87.27	90.99	1.90	2.86	0.63	94.55	87.09	90.75	2.13
S3	32.64	0.79	94.55	84.55	90.73	2.98	3.48	0.60	93.64	88.18	90.56	1.35
S4	32.52	0.81	95.45	88.18	91.60	2.03	3.03	0.60	94.55	87.27	91.35	1.73
S5	32.57	0.76	93.78	87.27	90.83	1.91	3.18	0.62	94.55	86.36	90.80	2.01
S6	35.99	0.81	94.55	89.09	91.37	1.40	3.29	0.60	94.55	86.36	91.36	2.14
S7	33.72	0.72	94.45	87.27	90.81	1.76	3.04	0.59	93.64	86.36	90.58	1.94
S8	34.79	0.80	93.64	87.27	91.10	1.90	3.39	0.59	94.55	87.27	90.93	1.82
S9	33.80	0.81	94.55	87.27	91.00	1.69	3.53	0.61	94.55	87.27	91.00	1.58
S10	35.02	0.80	94.55	87.27	90.54	2.15	3.51	0.60	94.55	84.55	90.23	2.78
S11	35.63	0.82	94.55	86.36	91.01	2.17	3.53	0.60	94.55	89.09	90.90	1.55
S12	35.62	0.80	93.64	87.00	90.75	1.95	3.27	0.62	93.64	85.45	90.70	2.15
S13	36.37	0.82	94.55	88.18	91.37	1.97	3.52	0.60	94.55	87.07	91.25	2.01
S14	35.72	0.83	95.45	90.91	92.50	1.44	3.51	0.60	95.44	87.27	91.44	2.19
S15	35.30	0.81	94.55	88.18	91.31	1.88	3.30	0.74	93.64	86.36	91.19	2.08
S16	36.21	0.77	94.55	88.18	91.18	1.67	3.94	0.63	93.64	86.36	90.96	2.02
S17	36.03	0.82	94.55	87.27	91.34	1.78	3.58	0.60	95.45	87.27	91.13	2.57
S18	35.85	0.83	94.55	89.09	91.23	1.78	3.56	0.60	94.55	87.27	91.04	1.93
S19	36.15	0.84	95.45	84.55	90.86	2.53	3.38	0.61	94.55	86.36	90.78	2.39
S20	33.85	0.75	96.36	85.35	91.11	3.09	3.40	0.60	94.55	85.55	90.73	2.56

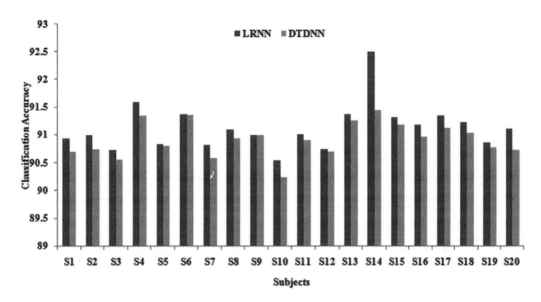

FIGURE 4.6
Classification accuracy for Parseval features using LRNN and DTDNN.

TABLE 4.3

Single trail analysis for Parseval features using LRNN

	Single Trail Analysis											
	Events								Nonevents			
Tasks	R	L	UR	DR	UL	DL	RM	LM	O	C	S	Unknown
S1	7	8	7	6	8	9	8	8	8	7	8	12
S2	6	10	6	6	7	6	7	8	6	7	7	15
S3	6	6	7	6	7	6	8	9	6	6	8	13
S4	6	9	5	5	6	6	7	8	7	6	7	14
S5	7	8	6	6	7	6	6	7	7	8	6	13
S6	5	9	9	7	8	8	6	8	8	8	6	6
S7	5	8	7	6	7	7	6	6	7	7	7	11
S8	6	9	10	9	7	7	10	9	8	9	9	4
S9	8	7	7	6	9	9	9	7	9	9	9	8
S10	8	6	9	9	9	8	6	6	8	8	9	9
S11	7	8	7	6	7	8	6	6	8	6	7	11
S12	6	9	6	7	9	9	5	5	8	7	8	8
S13	6	9	8	7	9	5	6	8	9	8	9	9
S14	7	9	9	6	6	8	5	8	8	9	8	9
S15	6	8	8	7	6	8	6	8	8	7	7	13
S16	5	8	8	7	7	7	6	7	8	8	9	9
S17	5	8	8	6	8	8	8	9	9	9	9	9
S18	6	7	7	7	6	9	6	8	8	8	8	12
S19	8	9	7	6	7	8	7	7	8	9	9	7
S20	6	9	6	7	6	10	6	8	9	7	8	11

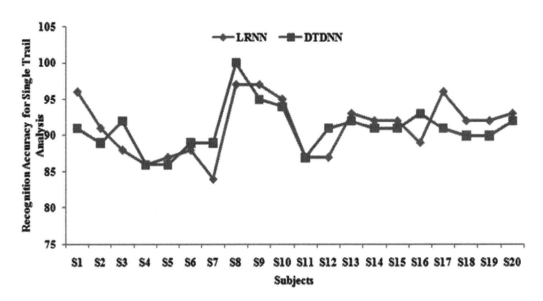

FIGURE 4.7

Single trail recognition for Parseval features using LRNN and DTDNN.

TABLE 4.4

Single trail analysis for Parseval features using DTDNN

	Single Trail Analysis											
	Events									Nonevents		
Tasks	R	L	UR	DR	UL	DL	RM	LM	O	C	S	Unknown
S1	6	7	8	6	8	9	8	8	7	7	7	10
S2	6	9	6	6	7	7	7	7	6	7	7	14
S3	6	8	7	7	7	6	7	9	6	6	8	15
S4	6	8	7	6	6	6	7	8	7	6	6	13
S5	7	7	6	6	7	6	7	7	6	7	6	14
S6	6	8	8	7	8	8	6	8	7	8	6	9
S7	6	8	7	6	7	7	7	7	7	8	6	13
S8	6	8	10	9	7	7	9	9	9	9	9	8
S9	7	7	7	6	9	9	9	6	8	9	7	11
S10	7	7	9	9	9	8	6	6	7	8	8	10
S11	7	8	7	6	7	8	7	6	8	7	7	9
S12	6	8	6	7	9	9	7	7	8	7	8	9
S13	7	8	8	7	9	6	6	8	8	8	8	9
S14	7	8	9	6	6	8	6	8	8	8	7	10
S15	7	8	8	7	7	8	6	8	8	8	7	9
S16	6	8	8	7	7	8	7	7	9	8	8	10
S17	5	8	7	6	8	8	8	9	8	8	9	7
S18	6	7	7	7	6	9	7	8	7	7	8	11
S19	6	9	6	6	7	8	7	7	8	8	8	10
S20	6	9	6	7	7	10	6	8	8	8	8	9

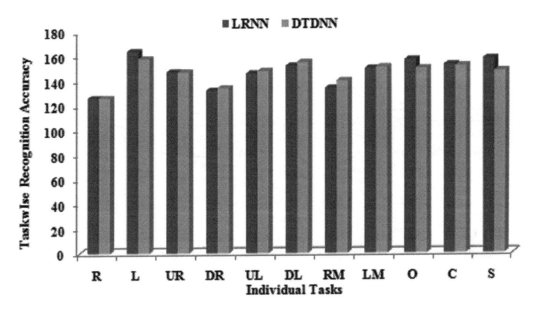

FIGURE 4.8
Task wise recognition accuracy for Parseval features using LRNN and DTDNN.

From Table 4.4 it is concluded that first maximum average recognition accuracy of above 92% is obtained by the subject S8, second maximum recognition accuracy of 84% is obtained by the subjects S9 and S10, minimum average recognition accuracy of 72% is achieved by subject S4, and the remaining subjects average recognition accuracy of 73% is achieved using Parseval features with DTDNN architectures is displayed in Figure 4.7.

From Table 4.3, we found individual first maximum task recognition of 82% for left, second maximum task recognition of 79% for stare, and minimum recognition of 63% is obtained for right using Parseval features with LRNN model. From Table4.4, we obtained individual first maximum task recognition of 79% for left, second maximum task recognition of 78% for down left, and minimum recognition of 63% is obtained for right using Parseval features with DTDNN model. Single trail analysis for twenty individual subjects for eleven tasks per ten trail results is shown in Figure 4.8.

From Tables 4.3 and 4.4, it is observed that the recognizing accuracy of the LRNN network model using Parseval features is comparatively better than that of DTDNN network models. From the eleven tasks, maximum recognition accuracy is obtained for the left task and minimum recognition accuracy obtained for right tasks. From the result it can be concluded that Parseval features with LRNN network model are more suitable for designing the nine states HCI.

4.6 Conclusion

This chapter analyze the possibility of designing nine states HCI system design using EOG. EOG signal data for eight events and three nonevents are collected from twenty subjects. Events and nonevents are assigned based on the voluntary and nonvoluntary eye movements possible by all subjects. Two new eye movements are proposed as events, namely, the rapid movement and the lateral movement in addition to the six movements used in the literature survey for designing nine states HCI. Signal features are extracted using the Parseval theorem. A layered recurrent neural network and distributed time delay network are used to identify the events and nonevents with a classification performance of 91.12% and 90.92%. The result obtained in this methodology is appreciable compared with the previous study [36]. Offline analysis of individual signals revealed the feasibility of designing the nine states HCI using EOG signals; however, verification of the system with real-time experimentation is required which will be the focus of our future research.

References

[1] J.K. Chacko, K.K. Deepu Oommen, N.S. Mathew, N. Babu. "Microcontroller Based EOG Guided Wheelchair," International Journal of Medical, Health, Pharmaceutical and Biomedical Engineering 7 no. 1 (2013): 482–485.

[2] S. Azam, A. Khan, M.S.H. Khiyal. "Design and Implementation of a Human Computer Interface Tracking System Based on Multiple Eye Features," Journal of Theoretical and Applied Information Technology 9 no. 2 (2009): 155–161.

[3] J. Tecce, J. Gips, C. Peter Oliveri, L.J. Pok, M.R. Consiglio. "Eye Movement Control of Computer Functions," International Journal of Psychophysiology 29 (1998): 319–325.

[4] A. Dix, J. Finlay, G. Abowd, R. Beale. Human-Computer Interaction, Prentice Hall, 2003.

[5] A. Bansal. "Design of an Electro-Ocular and Temperature Sensing Device for the Non-Invasive Health Monitoring System (NIHMS)," B.E. Thesis, McMaster University, Hamilton, Ontario, Canada, 2008.

[6] M. Mangaiyarkarasi, A. Geetha. "Cursor Control System Using Facial Expressions for Human-Computer Interaction," International Journal of Emerging Technology in Computer Science & Electronics 8 no. 1 (2014): 30–34.

[7] V. Aswin Raj, V. Karthik Raj. "EOG Based Low Cost Device for Controlling Home Appliances," International Journal of Innovative Research in Science, Engineering and Technology 3 no. 3 (2014): 708–711.

[8] R. Barea, L. Boquete, M. Mazo, E. Lopez. "System for Assisted Mobility Using Eye Movements Based on Electrooculography," IEEE Transactions on Neural Systems and Rehabilitation Engineering 10 no. 4 (2002): 209–218.

[9] S. Venkataramanan, P. Prabhat, S.R. Choudhury, H.B. Nemade, J.S. Sahambi. "Biomedical Instrumentation Based on Electrooculogram (EOG) Signal Processing and Application to a Hospital Alarm System," International Conference on Intelligent Sensing and Information Processing (2005): 535–540.

[10] A.A. Shaikh, D.K. Kumar, J. Gubbi. "Visual Speech Recognition Using Optical Flow and Support Vector Machines," International Journal of Computational Intelligence and Applications 10 (2011): 167–187.

[11] M. Jones, K. Grogg, J. Anschutz, R. Fierman. "A Sip-and-Puff Wireless Remote Control for the Apple iPod," Assistive Technology 20 no. 2 (2008): 107–110.

[12] A.B. Usakli, S. Gurkan, F. Aloise, G. Vecchiato, F. Babilon. "On the Use of Electrooculogram for Efficient Human Computer Interfaces," Computational Inteligence and NeuroScience 2010 (2009): 1–5.

[13] J. Keegan, E. Burke, J. Condron. "An Electrooculogram-based Binary Saccade Sequence Classification (BSSC) Technique for Augmentative Communication and Control," International Conference of the IEEE EMBS (2009): 2604–2607.

[14] C.D. Mello, S.D. Souza. "Design and Development of a Virtual Instrument for Bio-signal Acquisition and Processing Using LabVIEW," International Journal of Advanced Research in Electrical, Electronics and Instrumentation Engineering 1 no. 1 (2012): 1–9.

[15] L. Sun, S.-A. Wang, J.-H. Zhang, X.-H. Li. "Research on Electrooculography Classification Based on Multiple Features," International Journal of Digital Content Technology and Its Applications 6 no. 10 (2012): 35–42.

[16] T. Wissel, R. Palaniappan. "Considerations on Strategies to Improve EOG Signal Analysis," ACM International. Journal Artificial. Life Research l no. 2 (2011): 6–21.

[17] C.-C. Postelnicu, F. Girbacia, D. Talaba. "EOG-based Visual Navigation Interface Development," Expert Systems with Applications 39 (2012): 10857–10866.

[18] R. Barea, L. Boquete, S. Ortega, E. López, J.M. Rodríguez-Ascariz. "EOG-based Eye Movements Codification for Human Computer Interaction," Expert Systems with Applications 39 (2012): 2677–2683.

[19] A. Banerjee. "Electrooculogram Based Control Drive for Wheelchair Realized with Embedded Processors," M.E Thesis, Jadavpur University, Kolkata, 2012.

[20] M. Nakanishi, Y. Mitsukura, Y. Wang, Y.-T. Wang, T.-P. Jung. "Online Voluntary Eye Blink Detection Using Electrooculogram," International Symposium on Nonlinear Theory and Its Applications (2012): 114–117.

[21] D.S. Nathan, A.P. Vinod, K.P. Thomas. "An Electrooculogram Based Assistive Communication System with Improved Speed and Accuracy Using Multi-directional Eye Movements," International Conference on Telecommunications and Signal Processing (TSP) (2012): 554–558.

[22] A.S. Sherbeny, S. Badawy. "Eye Computer Interface (ECI) and Human Machine Interface Applications to Help Handicapped Persons," The Online Journal on Electronics and Electrical Engineering 5 (2013): 549–553.

[23] S. Ramkumar, K. SatheshKumar, G. Emayavaramban. "EOG Signal Classification Using Neural Network for Human Computer Interaction," International Journal of Computer Theory and Applications 9 no. 24 (2016): 1–11.

[24] C.R. Hema, M.P. Paulraj, A.H. Adom. "Improving Classification of EEG Signals for a Four-state Brain Machine Interface," IEEE-EMBS Conference on Biomedical Engineering and Sciences (2012): 615–620.

[25] N.P. Padhy. Artificial Neural Network, Artificial Intelligence and Intelligent Systems, New Delhi: Oxford University Press, p. 412, 2005.

[26] S.N. Sivanandam, S.N. Deepa. Principles of Soft Computing, Published by Wiley India (P) Ltd, 2007.

[27] G. Jialu, S. Ramkumar, G. Emayavaramban, M. Thilagaraj, V. Muneeswaran, M. Pallikonda Rajasekaran, A.F. Hussein. "Offline Analysis for Designing Electrooculogram Based Human Computer Interface Control for Paralyzed Patients," IEEE Access 6 (2018): 79151–79161.

[28] L. Junwei, S. Ramkumar, G. Emayavaramban, D. Franklin Vinod, M. Thilagaraj, V. Muneeswaran, M. Pallikonda Rajasekaran, V. Venkatraman, A.F. Hussein. "Brain Computer Interface for Neurodegenerative Person Using Electroencephalogram," IEEE Access 7 (2019): 2439–2452.

[29] S. Ramkumar, G. Emayavaramban, K. Sathesh Kumar, J. Macklin Abraham Navamani, K. Maheswari, P. Packia Amutha Priya. "Offline Study for Implementing Human Computer Interface for Elderly Paralyzed Patients Using Electrooculography and Neural Networks," International Journal of Intelligent Enterprise 6 (2019): 1–12.

[30] S. Ramkumar, G. Emayavaramban, K. Sathesh Kumar, J. MacklinAbraham Navamani, K. Maheswari, P. Packia Amutha Priya. "Task Identification System for Elderly Paralyzed Patients Using Electrooculography and Neural Networks," International Conference on Big Data Innovation for Sustainable Cognitive Computing, Springer Innovations in Communications and Computing (2018): 1-10.

[31] S. Fang, A.F. Hussein, S. Ramkumar, K.S. Dhanalakshmi, G. Emayavaramban. "Prospects of Electrooculography in Human-Computer Interface Based Neural Rehabilitation for Neural Repair Patients," IEEE Access 7 (2019): 25506–25515.

[32] C.R. Hema, S. Ramkumar, M.P. Paulraj. "Identifying Eye Movements Using Neural Networks for Human Computer Interaction," International Journal of Computer Application 105 no. 8 (2014): 18–26.

[33] C.R. Hema, M.P. Paulraj, S. Ramkumar. "Classification of Eye Movements Using Electrooculography and Neural Networks," International Journal of Human Computer Interaction 5 no. 3 (2014): 51–63.

[34] S. Ramkumar, K. Sathesh Kumar, G. Emayavaramban. "Nine States HCI Using Electrooculogram and Neural Networks," International Journal of Engineering and Technology 8 no. 6 (2017): 3056–3064.

[35] S. Ramkumar, K. Sathesh Kumar, T. Dhiliphan Rajkumar, M. Ilayaraja, K. Sankar. "A Review-classification of Electrooculogram Based Human Computer Interfaces," Biomedical Research 29 no. 6 (2018): 1078–1084.

[36] S. Ramkumar, K. Sathesh Kumar, G. Emayavaramban. "A Feasibility Study on Eye Movements Using Electrooculogram Based HCI," IEEE International Conference on Intelligent Sustainable Systems (2017): 384–388.

5

Smart Mobility System for Physically Challenged People

S. Sundaramahalingam, B. V. Manikandan, K. Banumalar, and S. Arockiaraj

Mepco Schlenk Engineering College (Autonomous), Sivakasi, Tamil Nadu, India

5.1 Introduction: Background

To meet the challenges for development in many countries, government has taken new initiative to implement smart cities, introduced special purpose vehicles, new development or retrofitting, etc. In our country, more than 30% of the population lives in cities which is nearly equal to the total population in the United States. A recent study states that the urban population will receive about 600 million in the year 2031. Also the number of metro cities is expected to be about 87 in the next 15 years. All this development of smart cities would require more investment in the infrastructure of cities. These funds will predict livability of the cities. In smart cities, vehicles are being developed by incorporating latest technology. There is no proper platform for physically challenged people that gives the idea to propose smart vehicles for them. This makes them to commute independently and they always depend on others to finish their travel. According to a study conducted by the Christober and Dana Reeve Foundation [1], about 1 in 60 million people has been diagnosed with paralysis. Physically challenged people use wheelchairs and require more physical strength to turn and steer wheels. This situation is more tragic in the case of the paralyzed and the quadriplegic. They always require some assistance but people are very busy in their own work and don't have sufficient time to spend with the physically disabled. In order to sort out this problem, a locomotive contrivance would serve a monumental purpose, especially one that comprehends and complies with the user's will. The challenge would thus be to devise and construct a locomotive contraption via a wheelchair which accords with the subjects' instinctive volition and reciprocates in a manner that will assist navigation. Davis et al. [2] proposed different hand gestures to control wheelchair. They have found eight types of hand gesture with greater accuracy to control it. Sourabh Kanwar [3] used USB camera to track eyeball movement controlled using LabVIEW. The idea here is to improve the independence of physically challenged people over complicated systems with complex functionalities, thus offering a locomotive device that comprehends the user's instinctive thought process enabling simplified motion [4–6]. Many researchers previously attempted search coil method, motion based and voice based method, etc., using various computer input devices like mouse and keyboard [7–10]. Many existing methods are too costly and are a burden to the user [11–15]. In this chapter, we develop a smart four wheeler chair vehicle fitted with two cameras in different locations. The cameras will be monitored by LabVIEW to make the user interaction easier. The eyeball and hand gesture are captured by the camera and the acquired images are processed by LabVIEW to start the control action based on the requirement. In LabVIEW, image processing using NI vision assistant and motor control is done using Linux MakerHub and Arduino ATMEGA 2560.

5.2 Methodology

Figure 5.1 shows the overall block diagram of this work. The eyeball movement and hand gesture are given as input to LabVIEW where image processing and motor control will be done. The output of motor control is given to Arduino ATMEGA 2560 which in turn is connected to ULN 2803 relay driver. The relay driver runs the motor accordingly.

5.3 Image Processing Module

The main goal of this project is to create an innovative solution without using expensive components. As compared to existing image processing boards with DSP chips, virtual instrumentation based image processing system is less expensive and less time consuming. We choose IMAQ Image processing module to determine the motion of eyeballs and hand. The image processing module gives images or parameters or characteristics as output. The relevant instructions are passed to the vehicle

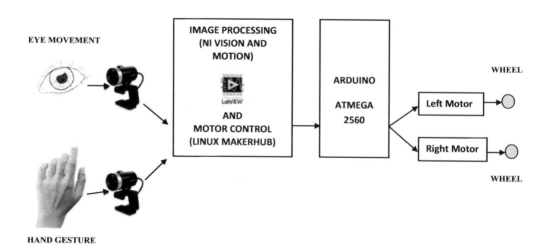

FIGURE 5.1
Overall block diagram of this work.

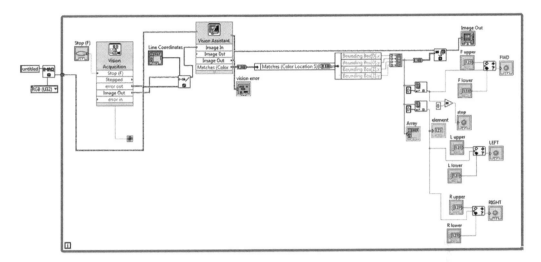

FIGURE 5.2
Front panel of image processing module in LabVIEW.

based on the movement of eye and presence of user's hand. IMAQ vision will analyze and filter the image. A web camera is connected with the head of a user to monitor eye movement and another camera is fitted in the vehicle to track user hand gesture. As shown in Figure 5.2, the input is taken from camera and temporary memory is created by IMAQ create VI which is given to NI vision assistant through NI vision acquisition. In vision assistant, color pattern is matched and the region of interest is overlaid by a bounding box using IMAQ overlay rectangular VI and the output is given to motor control. IMAQ set color pixel line VI changes a line of pixels from a color image. Depending on the image size, this VI obtains clusters with four 16 bit integers or array of 42 bit integers. The image received from NI vision acquisition VI is specified with pixel coordinates using an array. This image is processed and given to NI vision assistant VI.

5.4 Motor Control Module

5.4.1 Linux MakerHub

Using LabVIEW graphical user interface, it is easy to interface with familiar external equipments such as Arduino, NI myDAQ, NI compact RIO, and NI myRio. Also it is not much difficult to use some special purpose sensors including accelerometers, temperature sensors, and ultrasonic distance sensors by interfacing with LabVIEW. By using Linux MakerHub, data can be easily transferred to external embedded based controller. There are more than 100 built-in libraries available in LabVIEW to analyze the information received in LabVIEW. The information moves from embedded device to LINX through a serial port, USB, wireless, and Ethernet connection. This advanced feature of LabVIEW enables us to move the information quickly from an embedded device to

LabVIEW or vice versa without adjusting the communication, synchronization, or even a single line of C code. Motor control module VI is created using LabVIEW by choosing suitable palette from libraries as shown in Figure 5.3. It transfers the information received from image processing module to Arduino.

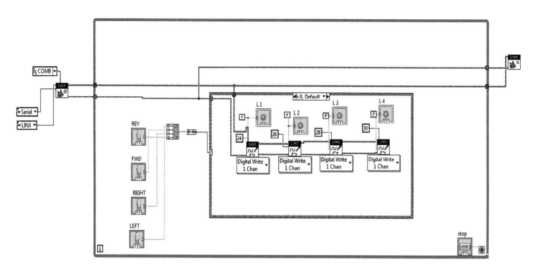

FIGURE 5.3
Front panel of motor control module in LabVIEW.

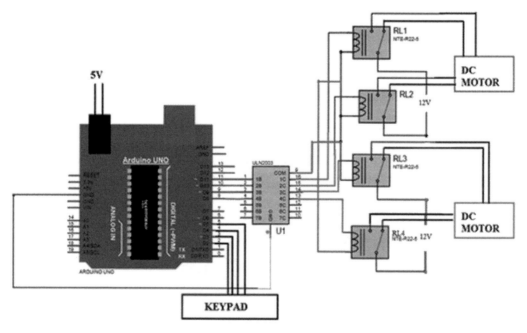

FIGURE 5.4
Control diagram of motor using Arduino.

5.4.2 Block Diagram of Motor Control

As shown in Figure 5.4, the input for the motor control is taken from switches which are later modified in such a way that the input sources are from image processing output. In motor control, five case structures were used for forward, right, left, stop, and default conditions. This output of motor control is given to Arduino ATMEGA 2560 which controls the operation of motor through relay. There are two 12 V DC gear motor with 100 rpm used in this project.

5.5 Results and Discussions

Eye movement detection and presence of human hand are captured by image processing control module. The LabVIEW output is interfaced with motor through Arduino and relay driver. It controls the movement of vehicle based on the eye position, that is, left, right, or forward movement. Once the direction of eyeball of user is sensed and processed to the controller (Arduino), it is used to control the DC motor as per instructions received from the image processing module. The program is written in Arduino in such a way that motor will be activated only if image of user hand is present otherwise motor will not operate as shown in Figures 5.7 and 5.8. It ensures security and safety of vehicle and user. The body of vehicle is designed with fiber glass to build lightweight vehicle. The suitable size of four wheels is attached with chassis, over that battery and electronic components are placed as shown in Figure 5.5.

FIGURE 5.5
Prototype model – Hardware setup.

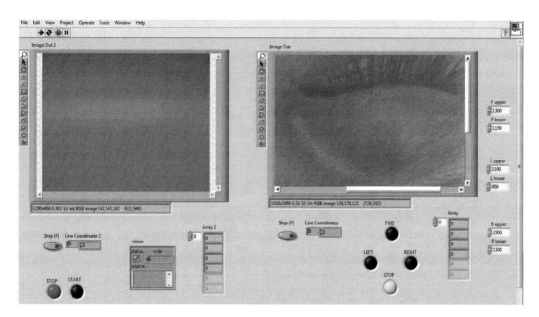

FIGURE 5.6

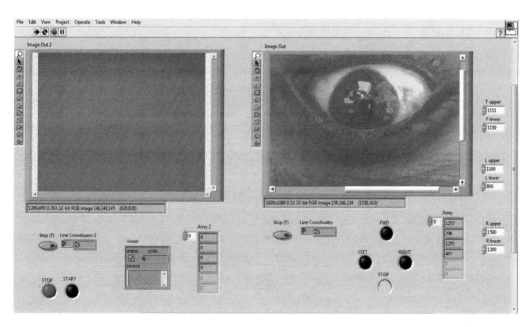

FIGURE 5.7

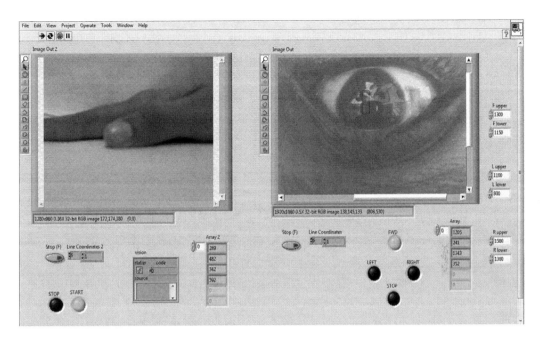

FIGURE 5.8

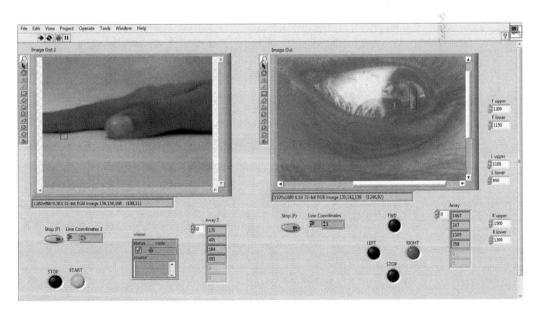

FIGURE 5.9

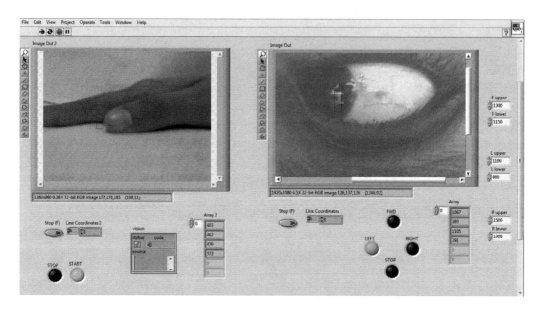

FIGURE 5.10

As shown in Figure 5.6, once the user's hand gesture and eyeball movement is detected by the camera, the controller receives the command to move forward. This signal enables the vehicle to move forward. If the user decides to change its moving direction, then he should move his eye toward the direction where he wants to go. Based on this instruction, the vehicle moves accordingly as shown in Figures 5.8–5.10. Further, the distance between wheelchair location and object is calculated, vehicle will move till targeted location is reached. The controller sends the stop command to the motor if smart wheelchair vehicle detects destination location. Then the motor will stop and the vehicle will not move.

5.6 Summary

Thus, smart mobility system for physically challenged people was designed and implemented using LabVIEW, Linux MakerHub, Arduino ATMEGA 2560, and other mandatory components. This enables to design cost-effective and efficient smart vehicle for physically challenged people who have no legs to travel easily without other's help. The system proved to be highly responsive to the user inputs and behaved precisely in the manner it was programmed to. The user's safety is most essential during his travel; optimum safety is designed in this system. In future, wireless control, obstacle detection, and various speed controls can be included in real-time application.

References

[1] R. Garg, N. Shriram, V. Gupta, V. Agrawal, "A Smart Mobility Solution for Physically Challenged," IEEE International Conference on Vehicular Electronics and Safety, pp. 168–173, 2009.

[2] J. Davis, M. Shah, "Recognizing Hand Gestures," Proceeding European Conference Computer Vision, Stockholm, pp. 331–340, 1994.

[3] R. Mardiyanto, K. Arai, "Eyes Based Electric Wheel Chair Control System," International Journal of Advanced Computer Science and Applications, 2, 12, 2011, 98–105.

[4] M. Selvaganapathy, N. Nishavithri, T. Manochandar, G. Manikannan, "Modern Vehicle for the Physically Challenged People Using Blue Eye Technology," International Journal of Mechanical Engineering and Technology, 2017, 208–212.

[5] D. Zhu, M. Khunin, T. Raphan, "Robust Hi-Speed Binocular 3D Eye Movement Tracking System Using Two Radii – Eye Model," International Conference of IEEE Engineering in Medicine and Biology, 2006.

[6] S. Kanwar, "Real Time Eye Tracking and Mouse Control for Physically Disabled," Conference ICL 2009, pp. 939–943, September 2009.

[7] C. Ma, W. Li, R. Gravina, G. Fortino, "Activity Recognition and Monitoring for Smart Wheelchair Users," IEEE 20th International Conference on Computer Supported Cooperative Work in Design (CSCWD), pp. 664–669, 2016.

[8] K. Miyawaki, D. Takahashi, "Investigation of Whole-Body Vibration of Passenger Sitting on Wheel Chair and of Passenger Sitting on Wheelchair Loaded on Lifter," International Symposium on Micro-NanoMechatronics and Human Science (MHS), pp. 1–6, 2016.

[9] N. Kobayashi, M. Nakagawa, "BCI-Based Control of Electric Wheelchair," IEEE 4th Global Conference on Consumer Electronics (GCCE), pp. 429–430, 2015.

[10] "Non-Invasive, Wireless and Universal Interface for the Control of Peripheral Devices by Means of Head Movements," Proceedings of the 2007 IEEE 10th International Conference on Rehabilitation Robotics, June 12–15.

[11] M. Shafi, P.W.H. Chung, "A Hybrid Method for Eyes Detection in Facial Images," International Journal of Electrical, Computer, and Systems Engineering, 231–236, 2009.

[12] R.C. Simpson, "Smart Wheelchairs: A Literature Review," Journal of Rehabilitation Resource Development, 2005, 423–436.

[13] L. Fehr, W. Edwin Langbein, S.B. Skaar, "Adequacy of Power Wheelchair Control Interfaces for Persons with Severe Disabilities: A Clinical Survey," Journal of Rehabilitation Resource Development, 37, 3, 2000, 353–360.

[14] R. Barea, L. Boquete, M. Mazo, E. Lopez, "System for Assisted Mobility Using Eye Movements Based on Electrooculography," IEEE Transaction on Neural System and Rehabilitation Engineering, 10, 4, 209–218, 2002.

[15] P.R.Singh, P.Shastry, J.D.Kini, F.Taheri, T.G.Giri Kumar, "Design and Development of a Data Glove for the Assistance of the Physically Challenged," International Journal of Electronics and Communication Engineering & Technology (IJECET), 4, 4, 2013, 36–41.

6

DHS

The Cognitive Companion for Assisted Living of the Elderly

R. Krithiga

Department of Electronics and Instrumentation Engineering, SRM Valliammai Engineering College, Affiliated to Anna University, Kancheepuram, Tamil Nadu, India

6.1 Introduction

Birth, childhood, adolescence, adulthood, and old age are the most important phases in one's life. All these phases have their own issues and troubles. Among these stages, the elderly phase is the most sensitive one. As each and every level passes, an individual's physical strength and mental stability deteriorate. Elderly people need more care, attention, and comfort to lead a healthy life without worries and anxiety. Lack of awareness regarding the changing behavioral patterns in elderly people leads to abuse of them by their kith and kin. Elders desire a life with good health, dignity, economic independence, and finally a peaceful demise. They long for care, love, and affection. Lending an emotional support to elders keeps them happy and jovial, which is inevitably the ideal way to live a healthy life. However, for many people, providing care and attention to elders is not possible due to their work priorities. Adults of this age are very busy with their own commitments and work pressure. This reflects on the time that they spend with elderly people at home and some adults fail to monitor the health needs of the elderly. There are various issues that govern the downfall of the health of older people. One of the main issues is negligence from their own family members.

The number of elderly people in the world is projected to be 1.4 billion in 2030 and 2.1 billion in 2050 and could rise to 3.2 billion in 2100. Globally, the number of persons aged 80 or over is projected to triple by 2050, from 137 million in 2017 to 425 million in 2050 [1].

This in turn creates the necessity of efficient caretakers of the elderly. This led to the development of assisted living robots that are capable of assisting elderly people in their daily routines like brushing, eating, bathing, walking from place to place, and so on [2–5]. In order to monitor their stress levels, loneliness, to track their medical performance, and also to remind them to take medicines timely, companion robots were introduced [6]. Even robots were developed as physical training assistants for the elderly in the work of P. T. V. Bhuvaneshwari, S. Vignesh, S. Papitha, R.S. Dharmarajan [3].

Both have equal pros and cons which are equalized by each other. Thereby, this chapter proposes the concept of DHS, which is an integration of these two types that can highly improve the standards of elderly care and be an optimal solution for taking care of their daily needs at their own homes. The basic idea of DHS is to bring happiness and composure in the lives of the elderly and to provide a solution for adults who face issues in taking care of the elderly at home.

The rest of the chapter is organized as follows: Section 6.2 discusses about assisted living robots. Section 6.3 discusses about companion robots and its types. The proposed method is discussed in Section 6.4. Finally, the conclusion and future work are stated in Section 6.5.

6.2 Assisted Living Robots

Assisted living is essential for people with physical challenges or for elderly people who cannot live independently. Elders need assistance for performing daily activities. There are many assisted living devices in our day-to-day life like an electric mobility scooter, freedom bed grip handle, lively wearable, uplift seat assist, activator poles, etc.

6.2.1 Electric Mobility Scooter

This battery-powered three or four-wheeler vehicle is designed for people who need help getting from place to place. This device is designed for both indoor and outdoor use. They are suitable for shopping trips to the grocery store or mall, or for long outdoor excursions. Though it is not a robot or autonomous vehicle, it assists the elderly people with their daily activities.

6.2.2 Care-O-Bot

As M. Hans, B. Graf, R.D. Schraft propose in [7], it is a robotic home assistant, which is a typical example of an assisted living robot. Care-o-bot can perform household tasks such as fetching and carrying objects, grasping, holding, lifting objects. It is also used as a walking aid. It can manage electronic devices at home and also acts as a daytime manager to remind the person to take medicines.

6.3 Companion Robots

"Artificial human companions" may be any kind of hardware or software designed to give companionship to a person. They can mentally assist the elderly and help them in maintaining an acceptable standard of life.

6.3.1 Buddy

Buddy the robot can ensure the well-being of senior citizens and can ease their loneliness at home by providing social assistance and interaction. Buddy can remind elders about upcoming events, appointments, and deliveries. It can detect falls and unusual activity and provide medication reminders. With buddy, seniors can access technologies, like Skype, Facebook, etc., with much more simplicity [8].

6.3.2 ElliQ

ElliQ is a tabletop robot that lights up when it hears its name. It does not have a face, arms, or legs, but it is voice activated. It talks and tries to keep its human companion active and engaged. The robot mimics head movements to connect with the user. ElliQ comes with a touchscreen tablet through which the user can interact and access the web and social media [9].

6.3.3 Animal-Assisted Therapy

There are also animal-assisted therapeutic robots, since pets such as cats and dogs exhibit a wide range of behaviors and emotions and help prevent depression in the elderly, though the therapeutic value of artificial pets remains inhibited by technology.

6.3.3.1 Paro

Paro is a therapeutic baby harp seal robot. As Takanori Shibata, the creator of Paro says in [6], it has touch-sensitive whiskers and a delicate system of motors and actuators that silently move its limbs and body. Paro is designed to actively seek out eye contact, respond to touch, remember faces, and respond to sounds. The robot responds to petting by moving its tail and opening and closing its eyes and even more interestingly, it can learn names. When humans interact with Paro, it perceives the sensory information from humans and generates its behavior in reaction.

6.3.3.2 Aibo

Aibo is adorned with artificial intelligence capabilities. It has facial recognition features, so it can detect different members of the family. It is designed to develop its own personality over time and it has received a positive response among people. It also has an increasing demand in the market. Aibo is extensively used for entertainment and also in elderly care around the globe.

6.4 Proposed Work: DHS Solution for All Needs in Elderly Care

DHS by the name expands as:

- D-Dexterous
- H-Humanoid
- S-Salubrious

Therefore, the name says that DHS is a humanoid robot which is dexterous, that is, skillful, quick, and salubrious, that is, which promotes health. DHS is designed especially for patients with cognitive impairments like Alzheimer's disease and dementia. These kinds of cognitive impairments mainly arise due to loneliness and depression. So, DHS aims to eradicate loneliness in the lives of elderly people and tries to keep them both physically and mentally active and also assists them in their daily routines. Simply stating, DHS is an innovative integration of both assisted living robots and companion robots.

DHS is a cognitive robot, that is, it has intelligence as that of humans. It can reason, remember, learn and communicate with humans too. As S. C. Mukhopadhyay, G. Sen Gupta describe in ref. [10], there are a set of subsystems, in which each of them are designed to do specific tasks and work together as an environment for elderly care. DHS has a system much more similar to it, but it integrates all the subsystems into one single entity as a humanoid robot.

The key idea behind DHS is to provide a solution to both the physical and mental issues of elderly people by assisting physically and providing a company mentally.

6.4.1 Features of DHS

DHS tries to imitate the way how humans perceive and react to things. DHS will have a human like face to express friendly gestures to elderly people at home, thereby improving human–robot interaction (HRI) as well as ensuring that the elderly person is not talking to a mere chatbot. There are six different senses in our body—vision, taste, smell, hearing, touch, and the mind. These differentiate humans from other mammals calling our race as "social animals." From the above-mentioned senses, DHS tries to incorporate the following senses.

6.4.1.1 Vision

DHS has two cameras in the place of eyes and it acts as the primary sensor to imbibe data regarding the environment. There is another camera in the back of its head providing the view behind it, to monitor the home completely. These cameras are also night vision cameras. Based on the camera inputs, DHS is capable of face recognition and object recognition. These environmental data are then sent to the internal computer which manipulates the information via digital image processing and then takes a decision in regard to the input. Camera sensor is always turned on for safety aspects so as to monitor 24×7 the elderly person's activities and the home. Apart from the camera, DHS also has four IR sensors in both hands and legs for obstacle detection.

6.4.1.2 Hearing

DHS has a total set of 10 microphones in its head to analyze from which direction the sound has come and to recognize the audio inputs precisely. The presence of 10 microphones is to mainly enhance speech recognition of DHS and to process multiple utterances at the same time. It is capable of distinguishing human voice, environmental sounds, and noise with a better efficiency. Based on the received inputs, DHS can also reply to the concerned person instantly by means of synthesized voice with those proper intonations expressing the exact emotion.

6.4.1.3 Touch

DHS incorporates tactile sensors in its arms and fingers so as to sense the objects and people. The most number of degrees of freedom (DOF) is incorporated in the fingers so as to completely understand and hold the object with perfect grip. DHS also has a suitable grasping force in order to handle objects safely. These touch sensors help to identify the presence or absence and also to analyze the shape and structure of an object. Touch sensors help DHS to react more like a human and improve HRI. It can give a handshake while meeting new people and use hand gestures while talking.

6.4.1.4 The Mind

The mind of the robot corresponds to the internal computer of the robot. It consists of an embedded computer whose software is designed using a suitable robotic operating system (ROS). It will also have a graphic processing unit (GPU) to process the data from the camera. The embedded computer will have an estimated high-speed processor in higher order GHz and possibly for memory it will either have cloud storage or a memory built within the robot. This system requires a processor with such a speed since it has to constantly manipulate inputs from all the sensors, take a decision instantaneously, therefore it has to process the information at a very high speed. It also requires large memory so as to store the sensory information.

6.4.2 Special Features of DHS

6.4.2.1 Walking Aid

In addition to its humanoid design, it also has a lever like structure in its arms, which will act as a walking aid to the elderly. The levers are present in both sides of the arms.

6.4.2.2 Wearable Hand Gear

DHS comes along with a wearable hand gear, which is specially designed to monitor physical parameters like temperature, pulse rate, etc. This is the only part of DHS which will be in continuous contact with the person. The gear has suitable sensors which will record the temperature and pulse rate from the person, and via the Internet it is sent to the robot. Thereby, DHS has a complete medical tracking of the person. The data are sampled by the gear for every 2 seconds.

6.4.2.3 Fall Detection Technology

DHS has fall detection technology. The wearable hand gear will also have a gyro sensor which detects an appreciable amount of vibration. In the absence of the robot, or if it is located far from the person, the sudden force exerted on the patient due to the fall is sensed by the hand gear and immediately sends information to DHS. In return the robot will inform the family doctor as a medical emergency and also sends a message to the family members.

6.4.2.4 Tablet

DHS also has a tablet attached to it through a lever which is foldable with three DOF. Using this tablet, DHS can display multiple information like reminders, medical analysis, important dates, etc. Using the tablet, the elderly person can play games, read books, view photos, listen to music, and even watch movies. People can also access different apps like Facebook, Instagram, etc. Therefore, DHS can provide even multimedia assistance and entertainment.

6.4.3 Design of DHS

DHS is a humanoid robot with 56 DOF. The DOF classification is as follows:

1. Neck = 2
2. Arms + fingers = 36
3. Hip = 3
4. Legs = 6
5. Levers for walking aid = 2
6. Support of the tablet = 3
7. Facial gestures = 4

Total: 56 DOF

The neck joint has 2 DOF for forward, backward and left, right movements. Both the arms have 6 DOF and fingers in total have 30 DOF. The hip joint has 3 DOF for roll, pitch, and yaw movements. Facial gestures are basically indicated by the two eyebrows and jaws which have 4 DOF. Therefore, with these structures, a humanoid robot can efficiently work for the specified design.

6.4.4 Cognitive Architecture of DHS

DHS is basically a cognitive companion, so it has the capability to think and learn from its experiences. DHS utilizes the concept of machine learning, deep learning, and adaptive learning. The proposed architecture shown in Figure 6.1 gives us the glimpse of how DHS perceives data and takes action in regard to it. The software for this architecture can be designed using suitable ROS.

The robot first collects information from the environment through its sensors, then low-level task recognition, which is to identify people and objects at an elementary level, is performed. Next, high-level task recognition, that is, analyzing the inputs, image processing, and speech processing is performed. This is then sent to the task planner, which collects information from the knowledge database which is either located in a local server or it is inbuilt in the robot.

The knowledge database also contains the basic commands and rules which have to be followed by the robot. The robot collects data from the knowledge base and the information regarding the task to be performed is collected. The tasks are arranged in a stack in the task arrangement module and it is made ready for execution in low-level task execution module. The task is finally executed by the actuators which are the high-level task executors. From each perception and execution of a task, the experiences are stored in the learning module, which influences the task planner in repeated tasks. Instead of fetching data from the knowledge base, the learning module provides information in regard to past experiences. This learning module can be accessed at a much faster rate since it is an integral element of the robot software and the memory size of the learning module is much less compared to the knowledge database.

By storing the experiences, the robot learns about a person or an object over a period of time, which enhances human–robot interaction and it helps to create a bond between the robot and the elderly person at home. This feature of DHS can reduce loneliness among elderly people to a great extent.

Robot software

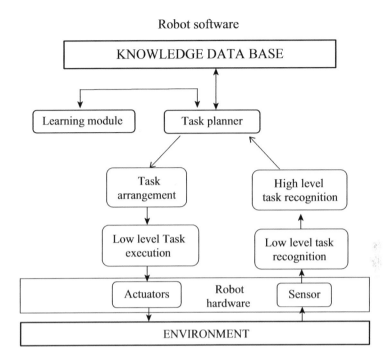

FIGURE 6.1
Cognitive architecture of DHS.

6.5 Conclusion and Future Work

The main reason for lack of attention toward elderly people is they have less contribution toward the society. They are neither a major part of the world's problems nor their solution. In a world like this, it is better to focus on personalized attention given to the elderly, rather than having them physically and letting them disappear mentally. Personalized approach proposed in this chapter will give a dogmatic solution for elderly care by supporting their daily activities and also lending a moral support. I hope soon senior citizens will also focus on their own happiness instead of worrying and getting depressed about being lonely. With lots of hope, I believe that the elder society will welcome DHS in their lives.

The only limitation of this work is it requires a very high-speed processor to act instantaneously based on the received data, but it can be implemented in the near future with the advent of IC technology. Building low power consumption systems and its design, and achieving dynamic balance for this design remain as the future work of this proposed approach.

References

[1] www.un.org/en/sections/issues-depth/ageing/
[2] Derek McColl, Wing-Yue Geoffrey Loie, Goldie Nejat, "Brian 2.1 a Socially Assistive Robot for the Elderly and Cognitively Impaired," IEEE Robotics and Automation Magazine, pp. 74–83, March 2013 edition.

[3] P. T. V. Bhuvaneshwari, S. Vignesh, S. Papitha, R.S. Dharmarajan, "Humanoid Robot Based Physiotherapeutic Assistive Trainer for Elderly Health Care," 2013 International Conference on Recent Trends in Information Technology (ICRTIT).

[4] Jessica P. M. Vital, Micael S. Couceiro, Nuno M. F. Ferreira, Carlos M. Figueirdo, Nuno M. M. Rodrigues, "Fostering the NAO Platform as an Elderly Care Robot," IEEE 2013.

[5] Roxana M. Agrigoraie, Adriana Tapus, "Developing a Healthcare Robot with Personalized Behaviors and Social Skills for the Elderly," 2016 IEEE.

[6] Takanori Shibata, "Therapeutic Seal Robot as Biofeedback Medical Device: Qualitative and Quantitative Evaluations of Robot Therapy in Dementia Care," Proceedings of the IEEE, Vol. 100, August, 2012.

[7] M. Hans, B. Graf, R.D. Schraft, "Robotic Home Assistant Care-o-bot: Past-Present-Future," Proceedings of the 2002 IEEE International Workshop on Robot and Human Interactive Communication, Berlin, Germany, 25–27 September 2002.

[8] www.bluefrogrobotics.com/en/buddy-elderly-senior-robot-alone/

[9] www.voanews.com/a/israeli-company-creates-anti-aging-robot/4201292.html

[10] S. C. Mukhopadhyay, G. Sen Gupta, "Sensors and Robotic Environment Care of the Elderly," ROSE 2007- IEEE International Workshop on Robotic and Sensor Environments, Ottawa, Canada, 12–13 October 2007

7

Raspberry Pi Based Cancer Cell Detection Using Segmentation Algorithm

S. Yogashri, S. Jayanthy, and C. Narendhar

Sri Ramakrishna Engineering College, Coimbatore, Tamil Nadu, India

7.1 Introduction

Cancer detection has always been a major issue for diagnosis and treatment planning. Most of the world's population gets cancer at some point during their lifetime. Cancer is one of the common diseases which is responsible for maximum mortality with about 0.3 million deaths per year. The chances of getting affected by this disease are due to lifestyle changes such as increase in use of tobacco, deterioration of dietary habits, lack of activities, and many more. The possibility of curing from cancer is increased due to recent advancement in both medicine and engineering. The chances of curing from cancer primarily depend on its detection and diagnosis. There are several techniques for detection of cancer. These techniques may include various imaging techniques such as X-ray, Computer Tomography (CT) Scan, Positron Emission Tomography (PET), Ultrasound, and Magnetic Resonance Imaging (MRI) and pathological images.

For accurate detection of cancer, pathologists use histopathology biopsy images for examining the microscopic tissue structure of the patient. Thus, biopsy image analysis is a vital technique for cancer detection. Histopathology is the study of symptoms and indications of the disease using electron microscopic biopsy images. The main aim of the proposed work is to design and develop a low-cost system for automated detection and classification of cancer using electron microscopic biopsy images.

Since image processing algorithms are used, time complexity can be reduced by implementing them in ARM Cortex processors. The proposed work is based on an embedded system consisting of Raspberry Pi 3 as a low-cost ARM powered Linux based computer. The ARM Cortex 64 bit embedded platform in the Raspberry Pi 3 supports floating point operations thus improving the real time performance of the system.

Section 7.2 describes the related works, Sections 7.3 and 7.4 describe the methods and models, Sections 7.5 and 7.6 describe the results and discussions, and Section 7.7 gives the conclusion of the work presented in this chapter.

7.2 Related Work

In the study of Kalaivani et al. (2017), the design of the system is ideal for performing a Computer Aided Detection system for detection of lung cancer. It can be helpful in assisting pathologists in the detection process by validating their diagnosis to prevent inaccurate diagnosis and also the system can perform even better if it is trained using larger databases and also be used for detecting lung cancer without the presence of pathologists in the future.

In the study of Meena Prakash et al. (2017), a method to segment the breast image using three segmentation techniques, K-Means, Fuzzy C Means and Gaussian Mixture Model Maximization is implemented. The conversion of color space is done to enhance the color analysis for classification of the images into benign and malignant stages. Breast thermography uses infrared camera to generate high-resolution images of various changes in the breast. Precancerous and cancerous tissues are highly proliferating tissues and they need large amount of nutrients. So they derive new blood vessels from the body which results in increase in surface temperature. Thermography is a technique used to detect cancer earlier than mammography technique.

In the study of Hazra et al. (2017), MRI images are used for detecting brain tumor. Various image processing techniques and their requirements and properties in the context of brain tumor detection using MRI scanned images are discussed in this chapter. The application of segmentation and edge detection is highly beneficial for medical diagnosis. To differentiate tumor affected regions from various brain tissues, thresholding segmentation technique is used. Using this algorithm, identification of brain tumor is done efficiently.

Hebli and Gupta (2017) have dealt with the features of neighboring double examples and gray level co-occurrences are removed from brain images with benign or malignant or normal images. In the training mode, the removed features are trained using PNN classifier and in classification mode the same features are extracted from test brain image using PNN classifier. When the test image is not same as any training image, the image can be included in training set data. By the comparison between PNN and CNN, PNN is considered to have major advantages. It is due to the fact that PNN learns from training data instantaneously and due to the speed of learning capability, it can adapt its learning in real time.

In the study of Firke and Phalak (2016), an image processing technique has been used to detect early stage lung cancer with CT scan images. The CT scan image is first preprocessed followed by segmentation of the region of interest (ROI) of the lung. Discrete waveform transform method is applied for image compression and then the features are extracted using a GLCM. After that the results are fed into an SVM classifier to determine if the lung image is cancerous or non-cancerous.

The study of Sudharsan et al. (2016), mainly focuses on detecting and localizing the brain tumor region by using patient's MRI images. This proposed methodology consists of three stages—preprocessing, edge detection and segmentation. Preprocessing stage involves converting original image into a gray scale image and it is followed by edge detection using Sobel, Prewitt, and Canny algorithms with image enhancement techniques. Next, segmentation is applied to clearly display the tumor-affected region in the

MRI images. Finally, the image is clustered using the k-means algorithm and MATLAB is used for the development of the project.

Warude and Singh (2016), have described a complete automatic computerized method for WBC identification, counting and classification using microscopic images. This technique can identify WBC's present in an image and can separate the grouped and ungrouped WBC followed by nucleus selection and feature extraction. Finally, counting and classification of leucoplast is done.

In the study of Dilip Kumar et al. (2015), an image processing is used to find the tumor from the MRI scanned images. The segmentation algorithm for detecting the boundary and extraction of tumor is done with the Simulink model using MATLAB. The segmentation is mainly used to extract various features and information.

In the study of Al-Mohair et al. (2015), an ANN-based processing approach is used to detect skin cancer by means of K means clustering algorithm. This methodology uses the following techniques: Data collection, preprocessing, segmentation, feature extraction and classification. The dermoscopy images are extracted from the 2D wavelet transformation method, then the values are given as input nodes to the neural network.

In the study of Murthy and Sadashivappa (2014), the thresholding and morphological operations of efficient brain tumor segmentation is carried out. This is the efficient algorithm where segmentation of tumor is carried out and its features such as centroid, perimeter and area are calculated from the segmented tumor. To detect brain tumor, scanned MRI images are given as input. The work involved here helps to detect tumor and its features help in giving the proper treatment plan to the patient.

In the proposed work, K-Nearest Neighbor matching algorithm has been used to design an automated process for early stage detection of lung cancer.

7.3 K-Nearest Neighbor Matching Algorithm

K-Nearest Neighbor Matching algorithm stores and classifies the data based on a similarity measure. It has been used in statistical estimation and pattern recognition. The selection of K will determine how well the data can be used to generalize the results of the algorithm. A large K value has reduces the variance due to noisy data. The KNN() function finds the K-Nearest Neighbors using Euclidean distance where K is a user-specified number. All the distances correspond to points in an n-dimensional feature space. Then each instance is being represented with a set of numerical attributes. The comparison is done by comparing feature vectors of different K-Nearest points. Figure 7.1 shows the K-Nearest Neighbor matching algorithm which includes preprocessing, feature matching, image segmentation and image classification.

7.3.1 Preprocessing

In this stage, the given input image is being processed with all the images in the given dataset and the images are being compared with each image in the dataset using pixel

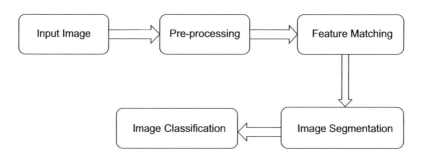

FIGURE 7.1
K-Nearest neighbor matching Algorithm.

by pixel comparison. It also makes predictions by calculating the similarity between an input sample and each training dataset.

7.3.2 Feature Matching

In this process, it uses descriptors to match the features of one image with the other image Equation (7.1) indicates the calculation of distance of the descriptors.

$$Dist = \sqrt{\sum_{i=1}^{m} (xi - yi)^2} \qquad (7.1)$$

7.3.3 Image Segmentation

In this phase, K-Nearest Neighbor algorithm forms clusters in order to segment the image and with the help of this cluster the needed portion of the image can be easily segmented.

7.3.4 Image Classification

The classification of images is one of the most challenging task for automatic detection of cancer cells. This phase will classify the image as benign, malignant, metastatic, or terminal. For image classification, K-Nearest Neighbor classifier is used here.

The steps involved when implementing the algorithm using Opencv is given as follows:

Step: 1 Determine the number of nearest neighbors, k.

Step: 2 Compute the distance between the training samples and the query record.

Step: 3 Calculate "d(x, xi)" i = 1, 2, …, n, where d denotes the Euclidean distance between the points.

Step: 4 Sort all training samples according to the distance values and arrange the calculated n Euclidean distances in non-decreasing order.

Step: 5 Use a majority vote for the class labels of K- Nearest Neighbors, and assign it as a prediction value of the query record.

Step: 6 Let k be a +ve integer, take the first k distances from this sorted list.

Step: 7 Find those k-points corresponding to these k-distances.

Step: 8 Let ki denote the number of points belonging to the ith class among k points.

Step: 9 If $k_i > k_j \ \forall \ i \neq j$, then put x in class i.

When using K-Nearest Neighbor matching algorithm in Opencv, it uses two matchers such as Brute-Force matcher and FLANN matcher. Each one has two parameters used for processing the given dataset.

The algorithm for BF matcher is as follows:

Step: 1 Compute the key points using SIFT descriptor

Step: 2 Assigning an orientation to the key points and generate SIFT features

Step: 3 Finds the K-Nearest Neighbors for each query descriptor

Step: 4 If crosscheck = False, finds the nearest neighbors for each descriptor and if crosscheck = True, the knn match method with k = 1 will only return pairs

Step: 5 Then return the consistent pairs

The algorithm for FLANN Matcher is as follows:

Step: 1 Detect the key points using SURF descriptor and compute the descriptor. The SURF descriptor is computed.

Step: 2 The key points are extracted based on the SURF threshold.

Step: 3 Calculate the maximum and minimum distances between the key points.

Step: 4 Match key point descriptors with the matcher and using KD-Tree.

Step: 5 Draw only good matches and show the detected matches.

7.4 Hardware Design

The proposed Embedded system consists of Raspberry Pi 3 board with ARM Cortex Processor for detection of cancer cells using the segmentation algorithm.

7.4.1 Block Diagram

The block diagram of the proposed system using machine learning algorithm is shown in Figure 7.2. The K-Nearest Neighbor matching algorithm is implemented in Opencv

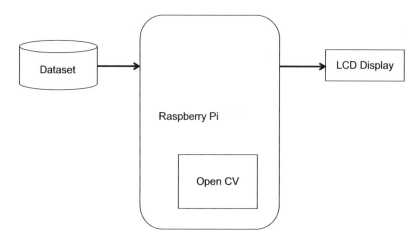

FIGURE 7.2
Block diagram of the proposed system.

Python. The algorithms are processed in the ARM Cortex processor. PC monitor is used to display the results indicating the various stages of the cancer cells. Figure 7.2 shows the block diagram of the proposed system.

7.5 Experimental Results

7.5.1 Experimental Setup

Raspberry Pi 3 board with ARM Cortex processor is the main processing unit. Power supply to the Raspberry Pi 3 board is through the USB cable from PC. The board can also be powered from a DC adaptor. The algorithms are executed by the ARM Cortex processor and the images are processed and the performance measure values are displayed in the LCD. These values are then processed to find the stages of cancer from the electron microscopy images. The output is displayed in the LCD.

7.5.2 Result Analysis

The original image is converted to gray scale image to reduce complexity. The noise in the gray scale image is removed using median filter and mean filter. Then the unwanted portions in the denoised image are removed and the features are extracted. The image is preprocessed, filtered, compressed, masked, segmented and extracted which results in image acquisition. The patches which represent the infected cancer cells are extracted and the output image is obtained in which the black spots indicate the presence of cancer infected cells. Figure 7.3 shows the stages involved in the detection of cancer cells.

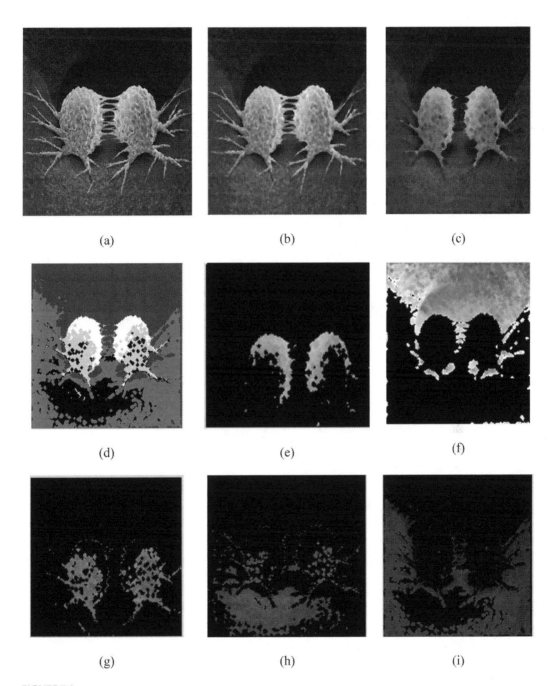

FIGURE 7.3
Stages involved in cancer detection: (a) Input image, (b) and (c) Filtering result using Weiner filter and Median filter, (d) Denoising result, (e) KNN result, (f) Mask generation, (g) and (h) Masking m1 and m2 results and (i) Final segmented cancer image.

7.6 Performance Measures

Around 16 datasets for different stages of cancer cells (Website 1) have been processed in the proposed work. The algorithms were implemented in both MATLAB and ARM processor. Various parameters have been measured and tabulated for four types of images. A benign tumor is not a cancerous tumor. Malignant tumor means the mass is cancerous. This tumor has the ability to multiply uncontrollably and metastasize to various parts of the body and invade surrounding tissue. Terminal is the advanced stage.

7.6.1 Accuracy

The measure of closeness of trained data to testing data is termed as accuracy. It is expressed in terms of percentage.

$$\text{Accuracy} = \frac{TP + TN}{N} \times 100\% \tag{7.2}$$

7.6.2 Sensitivity

Sensitivity is a measure of the proportion of positive samples which are correctly classified.

$$\text{Sensitivity} = \frac{TP}{TP + TN} \times 100\% \tag{7.3}$$

7.6.3 Specificity

Specificity is a measure of the proportion of negative samples that are correctly classified.

$$\text{Specificity} = \frac{TN}{TN + FP} \times 100\% \tag{7.4}$$

7.6.4 Packed Cell Volume

Packed cell volume (PCV) is the measure of the ratio of volume of each cell to the total volume of the whole cell.

$$\text{PCV} = \frac{\text{Volume of each cell}}{\text{Total volume of each cell}} \tag{7.5}$$

7.6.5 Tissue Density

Tissue density is the ratio of the mass of the cell by volume.

$$TD = \frac{\text{Mass of the cell}}{\text{Volume of the cell}} \qquad (7.6)$$

Accuracy is used to find the precision of detection of cancer cells. Hence, more accuracy means efficient detection of cancer cells at an early stage. During the benign stage a very good accuracy of 99% is obtained. Sensitivity finds the percentage of infected cells. Specificity finds the percentage of healthy cells from the infected cells. So greater sensitivity and specificity means there will be less error in distinguishing cancer cells from normal cells. Table 7.1 shows the performance measure values of the input images processed using MATLAB. Table 7.2 shows the performance measurement values of the input images processed using Open CV installed in ARM Cortex processor.

TABLE 7.1

Performance metrics in MATLAB

Stage	Accuracy (%)	Specificity (%)	Sensitivity (%)	PCV	TD	Execution Time (µs)
Benign	88.1390	90.1930	81.9922	0.3923	0.278	0.42
Benign	87.9856	90.1540	82.3124	0.3546	0.293	0.47
Malignant	82.4727	87.5432	80.1733	0.4321	0.428	0.56
Malignant	82.9564	86.9754	79.8765	0.4866	0.365	0.51
Metastatic	78.1763	81.6548	78.7092	0.5236	0.598	0.58
Metastatic	78.4521	82.0032	78.2456	0.5164	0.601	0.62
Terminal	75.1234	71.5401	67.5868	0.6356	0.622	0.67
Terminal	73.7895	71.8963	67.9741	0.6103	0.618	0.69

TABLE 7.2

Performance metrics in ARM processor

Stage	Accuracy (%)	Specificity (%)	Sensitivity (%)	PCV	TD	Execution Time (µs)
Benign	97.8546	98.7322	88.1132	0.4932	0.327	0.25
Benign	99.3261	97.3478	87.9785	0.4624	0.398	0.22
Malignant	94.3542	93.1457	86.7180	0.5289	0.488	0.37
Malignant	94.6023	92.9656	84.3674	0.5478	0.562	0.31
Metastatic	89.7956	87.6543	82.4960	0.5946	0.666	0.41
Metastatic	88.3981	87.1025	82.1043	0.6734	0.702	0.39
Terminal	79.3285	79.5872	70.1983	0.7212	0.782	0.43
Terminal	79.9910	80.1008	72.5123	0.7965	0.803	0.47

Packed cell volume and tissue density are measured to determine the extent of growth of cancer cell. Research has shown that the value of PCV along with other parameters indicates the extent of growth of cancer cells and the treatment to be given.

The cells having low density tend to produce a higher number of colonies (in case of lung cancer) than cells with relatively higher densities. It is also possible that low density cells have more proliferating ability than cells with higher density. Hence, it will be highly advantageous for the clinician as well as for the researcher if they could predict, on the basis of this parameter, whether the tumor will be metastasizing or not.

Comparing the performance values of algorithms implemented in MATLAB and ARM Cortex processor, the results obtained using the ARM Processor has higher accuracy, specificity and sensitivity values. The execution time is lesser when processed using ARM Processor.

Tables 7.3 and 7.4 show the comparative analysis values of the accuracy and specificity of input images processed using MATLAB and ARM processor.

Tables 7.5 and 7.6 show the comparative analysis values of the sensitivity and execution time of input images processed using MATLAB and ARM processor.

TABLE 7.3

Comparative analysis of Accuracy

Stage	MATLAB	ARM Processor	% Improvement over MATLAB
Benign	84.2760	97.0377	**12.7671**
Malignant	82.4727	94.3542	11.8815
Metastatic	78.1057	89.8754	11.7697
Terminal	69.2586	79.5089	10.2503

TABLE 7.4

Comparative analysis of Specificity

Stage	MATLAB	ARM Processor	% Improvement over MATLAB
Benign	90.1930	98.7322	8.5392
Malignant	87.5432	93.1457	5.6025
Metastatic	81.6548	87.6543	5.9995
Terminal	68.6257	78.5479	**9.9222**

TABLE 7.5

Comparative analysis of Sensitivity

Stage	MATLAB	ARM Processor	% Improvement over MATLAB
Benign	79.1279	88.6257	9.4978
Malignant	80.1733	86.7180	6.5447
Metastatic	75.6955	81.7456	6.0501
Terminal	66.6253	76.1660	**9.5407**

TABLE 7.6

Comparative analysis of Execution Time

Stage	MATLAB	ARM Processor	% Improvement over MATLAB
Benign	0.49	0.21	**0.28**
Malignant	0.55	0.31	0.24
Metastatic	0.64	0.40	0.24
Terminal	0.69	0.43	0.26

TABLE 7.7

Comparative analysis of Packed Cell Volume

Stage	MATLAB	ARM Processor	% Improvement over MATLAB
Benign	0.2350	0.5014	**0.2664**
Malignant	0.3181	0.5345	0.2164
Metastatic	0.5321	0.6947	0.1626
Terminal	0.6356	0.7965	0.1609

TABLE 7.8

Comparative analysis of Tissue Density

Stage	MATLAB	ARM Processor	% Improvement over MATLAB
Benign	0.186	0.331	0.144
Malignant	0.362	0.583	**0.221**
Metastatic	0.584	0.729	0.145
Terminal	0.622	0.803	0.181

Tables 7.7 and 7.8 show the comparative analysis values of the packed cell volume and tissue density of input images processed using MATLAB and ARM processor.

From the comparative analysis, it is inferred that benign stage cancer gets better accuracy, execution time and packed cell volume. The terminal stage cancer gets better specificity and sensitivity. The tissue density is better for malignant stage cancer. So from the comparative analysis the benign stage cancer gets better results when compared to the other three stages of cancer.

7.7 Conclusion

In this chapter, a method of detection of cancer cells using K-Nearest Neighbor matching algorithm, a simple machine learning algorithm is proposed for detection of cancer cells. If more datasets are used, the error rate in detection of cancer cells can be reduced.

The performance of the algorithm can be significantly improved using statistical machine learning algorithms for metric learning.

References

Al-Mohair, H. K., J. M. Saleh, and S. A. Suandi, Hybrid Human Skin Detection Using Neural Network and K-Means Clustering Technique, Applied Soft Computing, Vol. 33, No. C, (2015): 337–347, August.

Dilip Kumar, D., S. Vandana, K. Sakhti Priya, and S. Jeneeth Subhashini, Brain Tumor Image Segmentation Using MATLAB, IJIRST, Vol. 1, No. 12, (2015): 447–451.

Firke, O. K. and H. S. Phalak, Brain Tumor Detection Using CT Scan Images, IJESC, Vol. 6, No. 8, (2016): 2568–2570.

Hazra, A., A. Dey, S. K. Gupta, and A. Ansari, Brain Tumor Detection Based on Segmentation Using MATLAB, International Conference on Energy, Data Analytics and Soft Computing, Vol. 6, (2017): 425–430.

Hebli, A. P. and S. Gupta, Brain Tumor Detection Using Image Processing: A Survey, International Journal of Industrial Electronics and Electrical Engineering, Vol. 5, No. 1, (2017): 41–44.

Kalaivani, S., P. Chatterjee, S. Juyal, and R. Gupta, Lung Cancer Detection Using Digital Image Processing and Artificial Neural Networks, International Conference on Electronics Communication and Aerospace Technology, (2017):100–103.

Meena Prakash, R., K. Bhuvaneshwari, M. Divya, and K. Jamuna Sri, Segmentation of Thermal Breast Images Using K-Means Algorithm for Breast Cancer Detection, International Conference on Innovations in Information, Embedded and Communication Systems, (2017). DOI: 10.1109/ICIIECS.2017.8276142

Murthy, T. S. D. and G. Sadashivappa, Brain Tumor Segmentation Using Thresholding, Morphological Operations and Extraction of Features of Tumor, International Conference on Advances in Electronics Computers and Communications, (2014).

Sudharsan, M., S. R. Thangadurai Rajapandiyan, and P. U. Ilavarsi, Brain Tumor Detection by Image Processing Using MATLAB, Middle-East Journal of Scientific Research, Vol. 24, (2016): 143–148.

Warude, D. and R. Singh, Automatic Detection Method of Leukaemia by Using Segmentation Method, In International Journal of Advanced Research in Computer and Communication Engineering, Vol. 5, (2016): 495–498.

Website 1: www.canstockphoto.com/images-photos/cancer-cell.html

8

An AAC Communication Device for Patients with Total Paralysis

Oshin R. Jacob and Sundar G. Naveen

Karunya Institute of Technology and Sciences, Karunya Nagar, Coimbatore, Tamil Nadu, India

8.1 Introduction

Alternative and Augmentative Communication (AAC) is the term used to define various ways of communication other than speech, especially used by people with physical impairments. AAC has become the need of the century as 1.4% of the world's population suffer from developmental disorders.[1] The difficulty faced by the patients to convey his/her basic needs is a major challenge seen among this population. Often it requires the caretaker to intuitively understand the feelings and requirements of the patient and tend to them.[2] Of these, communication difficulties are extreme in case of tetraplegic and quadriplegic patients. Therefore, in order to design a device which is marketable among these sections of patients, it is important to understand the psychological mindset of the rural sect with respect to communication disabilities. One of the major parameters which must be considered in order to make a successful AAC device marketable (other than affordability) is the mortality rate of people associated with disorders relating to speech impairment. This is because mortality rate is inversely proportional to the demand of AAC devices among poorer sections of the society. High mortality rates are observed in cases associated with total paralysis, such as amyotrophic lateral sclerosis, cerebral palsy, and stroke, especially in poor countries, made worse by bad health conditions.[3] Popular AAC devices available for such patients are brain computer interfaces (BCI) and eye trackers which are costly, making these devices a priority than a necessity among the poorer sections of the society.[4] BCIs are still in their infancy in terms of development, and eye trackers involve complex computational algorithms, both of which directly contribute to the cost of the device. Cheaper devices like communication boards show inefficiency among tetraplegic patients. Thus, it is an emerging need to develop a communication device that is cheap, portable, available, and affordable. This chapter discusses such a device that uses blow from the patient to select predefined sentences on the device, which then gets communicated to the caretaker as speech.

8.2 Related Work

The market has seen a huge explosion of alternative and augmentative communication devices in this century. They range from brain computer interfaces, eyetrackers, and iPad apps[5] which are costly to more affordable communication boards. These devices have well-utilized the functioning parts of the human body to provide an effective means of communication. The common functions that a quadriplegic can exercise are the movement of eyes, breathing and blowing of air. Blowing air and the use of breath to control things has been an innovative field, and its application in the field of assistive technology has strengthened in the recent past. While breathing and blowing air takes effort, developing such a device would also be economical. The related devices in existence today include android applications using breath to communicate[6] and the use of Morse codes to form words.[7] But the challenge of practicability remains to develop a device that takes almost no effort from the patient to communicate his/her basic needs. The major limitations found in breath-enabled communication devices today is that it requires two inputs from the user, a long exhale and a short one to communicate. Similarly, blowing air uses the sip and puff method. This is tasking for a tetraplegic patient whose lungs are weak. Since two inputs are required by the user to enable communication, Morse codes are used to translate bits to letters. Hence, the time taken by the person to communicate even his basic needs is high.[8] This chapter discusses the development of a device which overcomes the above-mentioned challenges, and is economical, portable, and available to all. The chapter is organized as follows: Section 8.3 discusses the proposed system, Section 8.4 discusses the test results, and Section 8.5 is the conclusion.

8.3 Proposed System

8.3.1 Components Required

The system developed uses blow of air to act as a cursor to select a set of predefined options and output them as speech. The device uses a quartz crystal timer to select between a list of options on the board. The major components used in the making of this device include a peripheral interface controller (PIC), a Hall effect sensor, recording modules, a quartz crystal timer, four LEDs, and a speaker to output the selected sentence from a list of predefined sentences. The device also contains the "settings" option which is controlled by a 4 × 4 matrix switch and a 16 × 2 LCD display. The PIC18F45K22 controls the working of the speaker, the recording modules, the Hall effect sensor, the potentiometer, the quartz crystal timers, and the LEDs. Figures 8.1 and 8.2 show the block diagram of the PIC184F5K22 microcontroller and the components associated with it.

The components include the Hall effect sensor, four ISD1820 recording modules, speaker, 16 × 2 LCD display, the 4 × 4 matrix switch, and the reset button. The matrix switch is used to edit the settings displayed on the 16 × 2 LCD. The editable options included in this device are:

1. Ordering of the recorded modules with the sentence displayed as text.
2. Editing, updating, and deleting the predefined sentences stored and their recordings.
3. Setting an appropriate value for the timer. The options included are: 5 s, 10 s, 15 s, and 20 s.

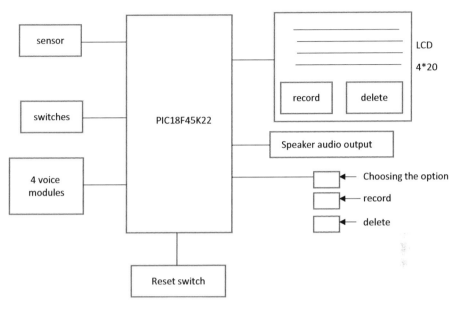

FIGURE 8.1
PIC18F45K22 and the components that it controls.

8.3.2 Working Principle

The device requires the patient to blow into a plastic tube, which activates the Hall effect sensor to sense the blow. A tiny fan is placed near the Hall effect sensor, thereby causing electric impulses to be produced from low air pressure given by the patient. The first blow from the patient starts the timer for 15 s. If no further blow is given by the patient within the prescribed time limit, the first option is selected as the speech output. For the patient to select consecutive options, the next blow must be detected within 15 s. Table 8.1 shows the list of editable options that is operable on the device.

The "settings" option: The caretaker uses this functionality to feed new sentences or make new recordings into the device as per the needs of the patient. Another important component in this device is the 4 MHz quartz crystal timer that is used to indicate the time of blow and the selection of options. It also gives the caretaker the flexibility to order the sentences and recordings according to the priority of the patient. The timer can also be set by the caretaker, depending on the condition of the patient. The time between the blows and selection of options can be set by the caretaker, and it ranges from 5 s to 20 s. The default value for the timer is 15 s. Providing this flexibility is important because of the weakness of the patient's lungs, which is indirectly affected by chest muscles, diaphragm, and abdomen which lacks strength in such cases. Table 8.2 shows the algorithm of the working of the device and Figure 8.3 indicates the flowchart of the working of the HALECOM device.

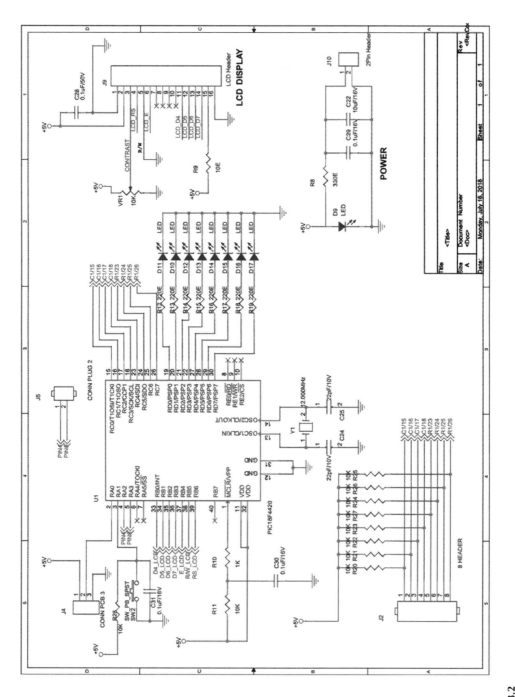

FIGURE 8.2
Detailed description of the microprocessor and its components.

TABLE 8.1

Default sentences that can be communicated

Option no.	Color of LED	Predefined sentences
1	Red	I am in pain
2	Green	I need to use the commode
3	Yellow	I am thirsty
4	Blue	I am hungry

TABLE 8.2

Algorithm used in the device

1. Initialize variable i = 0, set timer to 15s
2. Blow air into the device. i++, start timer
3. If (blow within 15s) then increment the value of i and go to step 3. Otherwise, option i is selected and the sentence in i is converted to speech.
4. End algorithm

8.3.3 Advantages and Applications

The advantages of the proposed system are:

1. User-friendliness: The device has no complications of using matrices or Morse codes. The approach followed for communication is straightforward, and the working is easy to understand.
2. Compatibility: The device can accommodate any language, as it has a recording module. The sensor used to detect blow from the patient is very sensitive and can detect very low pressures, which makes it comfortable for patients to use.
3. Flexibility: The speed with which sentences can be communicated can be adjusted by the caretaker depending on the condition of the patient. The timer range can be adjusted between 5 s and 20 s. The minimum time that can be set to communicate emergency sentence is 5 s.
4. Economical: This device is very economical and costs around 1000 rupees, which is the cheapest alternative and augmentative communication device in the world.

The device finds its use in various applications:

1. Communication device for people with strokes, spinal cord injuries, amyotrophic lateral sclerosis, cerebral palsy, tetraplegic patients, etc.
2. Communication device for patients in surgical wards after a tracheotomy operation, accident victims, etc.
3. AAC device for people suffering from total paralysis in rural areas.

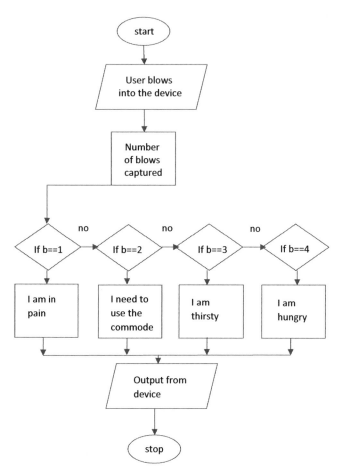

FIGURE 8.3
Flowchart of the working principle.

8.4 Tests and Results

The device was tested on seven patients in SEESHA Hospital, Coimbatore, India. The test results show that it requires a minimum difference of 40 ms between blows given by the patient into the device and the desired option to be selected. It is also observed that the minimum volumetric air flow rate required for the Hall effect sensor to activate is 2.0 L/min. Table 8.3 shows the air flow rate exerted by seven patients to communicate through the developed device for the list of four given options. Figure 8.4 graphically plots the values tabulated in Table 8.3.

In the graph shown, the "options" are the four editable predefined sentences mentioned in Table 8.1.

TABLE 8.3

Description of the volumetric air flow rate of seven patients in L/min for each of the four options listed on the device

Patient no.	Option 1	Option 2	Option 3	Option 4
1	4.26	4.11	4.23	4.3
2	3.23	3.24	3.1	3.3
3	4.2	4	3.76	3.8
4	2.12	1.98	2.13	2.11
5	3.13	3.24	3.1	2.76
6	2.21	2.12	1.89	2
7	3.45	3.89	3.76	3.34

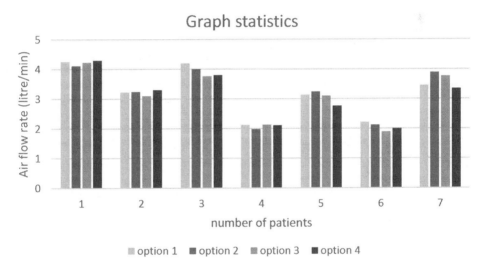

FIGURE 8.4
Graphical representation of the air flow rate of seven patients for a list of four options.

8.5 Conclusion

An alternative and augmentative communication device which is highly economical and user-friendly is developed to help people suffering from total paralysis communicate through subtle blows given by patient into a plastic tube.

References

1. Priyadarshi, Brajesh, and B. V. M. Mahesh. "Neurodevelopmental Disorders from a Clinical Linguistics Perspective." Emerging Trends in the Diagnosis and Intervention of Neurodevelopmental Disorders, 85–101. Accessed 24 February 2019. doi:10.4018/978-1-5225-7004-2.ch005.

2. Kroll, Thilo, and Melinda Neri. "Experiences with Care Co-ordination among People with Cerebral Palsy, Multiple Sclerosis, or Spinal Cord Injury." *Disability and Rehabilitation* 25, no. 19 (2003): 1106–1114. doi:10.1080/0963828031000152002

3. "Correction to: Heart Disease and Stroke Statistics—2018 Update: A Report from the American Heart Association." *Circulation* 137, no. 12 (2018). doi:10.1161/cir.0000000000000573.

4. Alper, Meryl. "Augmentative, Alternative, and Assistive: Reimagining the History of Mobile Computing and Disability." *IEEE Annals of the History of Computing* 37, no. 1 (2015): 96. doi:10.1109/mahc.2015.3

5. Hasselbring, Ted S., and Candyce H. Williams Glaser. "Use of Computer Technology to Help Students with Special Needs." *The Future of Children* 10, no. 2 (2000): 102. doi:10.2307/1602691

6. Gaba, Vinay. "Shwas: A Smartphone Based Augmentative and Alternative Communication (AAC) System Which Converts Breath into Speech." Inpressco.Com, 2014. https://inpressco.com/wp-content/uploads/2014/11/Paper83830-3832.pdf.

7. Kumar, S. Santhosh, B. K. Aishwarya, K. N. Bhanutheja et al., "Breath to Speech Communication with Fall Detection for Elder/Patient with Take Care Analytics." 2016 IEEE International Conference on Recent Trends in Electronics, Information & Communication Technology (RTEICT), 2016. doi:10.1109/rteict.2016.7807877.

8. Garcia, Ramon G., Joseph Bryan G. Ibarra, Charmaine C. Paglinawan et al., "Wearable Augmentative and Alternative Communication Device for Paralysis Victims Using Brute Force Algorithm for Pattern Recognition." 2017 IEEE 9Th International Conference on Humanoid, Nanotechnology, Information Technology, Communication and Control, Environment and Management (HNICEM), 2017. doi:10.1109/hnicem.2017.8269554.

9

Case Studies on Medical Diagnosis Using Soft Computing Techniques

Mary X. Anitha and Lina Rose
Karunya Institute of Technology and Sciences, Coimbatore, Tamil Nadu, India

Aldrin Karunaharan
International Maritime College, Liwa, Oman

Anand Pushparaj J.
NIT, Trichy, Tiruchirappalli, Tamil Nadu, India

9.1 Introduction

A major problem in image processing is the complexity due to unknown factors, human interference, ambiguous symptoms, etc. These issues can be solved by various soft computing techniques for image enhancement and feature extraction and classification of diseases. The problem lies in high accuracy and bulk of medical records which have to be minimized to overcome the errors during diagnosis to achieve high performance. Two case studies on diagnosis have been analyzed. The research illustrates that image processing techniques applications in cognitive technology have found breakthroughs in diagnostic research to a vast extent.

9.2 Case Study 1: Detection of Heart Failure Parameters Using 2D Echocardiographic Images Using Artificial Neural Networks

The economic development of a country is not praise worthy if the morbidity and mortality rates are increasing due to chronic heart diseases. The study shows the mortality rate due to heart failure is devastating since the past several years. Therefore, coronary heart failure is considered as a major concern in the medical field. Among several methods to diagnose heart disease, echocardiography seems to be the physician's choice. Moreover, a noninvasive technique is preferred worldwide for the patients' comfort. A detailed inspection of four heart chambers or cavities, together with pulmonary artery and vein and the heart valves, is done using electrocardiography (ECG). The analysis of visual images of the heart and the cardiac cycle helps in assessing myocardial health conditions.

Tremendous research and development are carried out for the detection of systole and diastole for various noninvasive diagnosis of cardiac disorders. The diastolic and systolic volume calculation involves the automatic detection of cardiac phases. However, depending on the image features, automatic and semiautomatic analysis can be done from electrocardiography. The work involves developing a new algorithm which estimates the cardiac muscular activity in electrocardiographic images that neither uses manual calculations nor segmentation procedures. This includes the prissiness of the algorithm to correctly detect atrial diastole and atrial systole. The approach is performed in various scenarios and classified and compared for better outcomes.

The analysis required different data samples of images acquired during separate stages of cardiac cycle which was approved by a radiologist. Image labeling was done by the radiology, which seems to be time consuming and involved much labor. From the literature, it was found that cognitive techniques such as artificial neural networks (ANN) would precisely distinguish systole and diastole phases. A faster result is obtained when the neural network training is accurately done by inputting more number of samples for several iterations. Hence, the trained network provides accurate results consuming time and labor effectively.

A faster detection can be achieved if ANN is trained well. This approach is time consuming and user friendly with less information to process. In addition, the most significant reason to develop such an algorithm is not limited to a research, but in this era, patients look for an affordable, efficient, and a portable device. A new possible approach in the field of heart disease remote monitoring could offer patients more individually focused care and, thus, an improved quality of life.

9.3 Heart Physiology

The heart is an unconquered paradigm of healthcare since ages. The general physiological characteristics of heart can be divided into two states working alternatively. The pumping of the heart and the heart rate are the two important parameters for measuring blood pressure, pulse, cholesterol, blood glucose, and any related ischemia.

The blood pumped to the aorta from the left cardiac chamber is then carried to the arterial circuit. The blood flow has varying pressure profile due to the load resistance offered by capillaries and arterials walls. Then, the right cardiac chamber pumps this blood to the pulmonary artery for gas purification and exchange at a lower pressure. Thus, the heart serves as a source of supplying blood to both the circuits simultaneously at varying flow rate and volume depending on the pressure. Each cardiac ejection is a result of maximum pressure at the bottom of the heart exerted to push blood upstream through the aorta, which is referred to as the systolic pressure, continued by a fall in pressure at the end of ventricular relaxation, which is referred to as the diastolic pressure. The former refers to the amount of pressure in the arteries during contraction of the heart muscles and the latter is the blood pressure when the heart rests between beats.

Though the primary concern is related to the mean pressure, the role of heart valves cannot be neglected when dealing with heart diseases, as it controls the blood flow to and from the heart. The cardiovascular system acts as a hydraulic system in regulation of flow, pressure, and other parameters. For every heartbeat, blood returning from the body and the lungs fills the atrial chambers and thereafter flow to the ventricular chamber by the controlled action of mitral and tricuspid valve at the opening of atria to ventricle. The valves shut the blood flow after a brief delay to ensure that no blood flows backward to atria. The ventricular contraction initiates the blood pumping through pulmonary and aortic valves. The former enhance the flow to the pulmonary artery from the right ventricle. Meanwhile, the latter allow the oxygen-rich blood to flow into the aorta from left ventricle, further carried to the body. The ventricular relaxation results in closing of pulmonary and aortic valves to restrict blood flow back into ventricular chamber. One of the main causes of varying pressure profile might be the improper functioning of the heart valves.

9.4 Methodology

9.4.1 Ultrasound Image Database Creation

The image database was created by collecting images from hospital. A sample size of 20 was fixed and the images corresponding to an individual with a normal heart and a deceased cardiac patient were collected. The cardiac cycle phase estimation is performed in apical two-chamber long-axis 0° view (LAX0) of 2D echocardiographic images. The image format was converted to portable network graphics (png) format. Figure 9.1 shows methodology of pressure measurement using 2D ECG image.

9.4.2 Image Preprocessing

Image resizing, image enhancement via filtering and grayscale, and binarization constitute image preprocessing. The raw images are reduced and thus the resized images make the subsequent processing faster. A grayscale image carries only intensity information of an image, that is, the value of each pixel is a single sample.

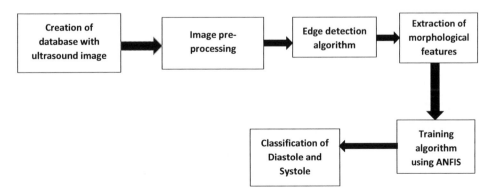

FIGURE 9.1
Methodology of Pressure measurement using 2D-ECG images.

The noise reduction is done using a median filter which is a typical preprocessing step to improve the results of later processing. The ability in preserving edges during noise reduction enables median filter to be used in almost all image processing applications. Apart from filtering, the images need to be visually clear enough to be analyzed, only then an image enhancement is done. The salient features of the original image are emphasized that will simplify the processing task which is accomplished using image enhancement techniques. An approach of replacing pixels with the mean of the neighboring values is conventionally carried out. Instead it replaces the median of those neighbor mean values. The surrounding neighboring pixel values are sorted numerically and replacing it with the middle value gives the mean value of the pixel.

9.4.3 Canny Edge Detection

This kind of detection algorithm is used to detect wide range of edges in various images usually known as optimal detector. The algorithm is intended to satisfy three necessary criteria—good detection, good localization, and minimal response. The Canny edge detection algorithm runs in five separate steps:

1. Smoothening (noise removal using blurring)
2. To detect images with large gradients
3. Marking local maxima as edges non-maximum suppression
4. Determination of potential edges using double thresholding
5. Narrowing all edges to a single edge to find the final edge (edge tracking).

9.5 Feature Extraction

The object's boundary consists of image pixels with high gradient which forms the boundary tracking. The follower to this pixel having the next highest gradient is determined using a 3 by 3 matrix scanning probe tracking, near the original pixel. Likewise, all pixels with high gradients are analyzed and the boundary is tracked. This results in the formation of closed contour. A question arises if there are two or more pixels with same gradient. The selection is based on the choice of the user and the number of pixels exhibiting the same value. Regionprops measures a set of properties for each connected component in the 2D image.

9.6 Artificial Neural Network Training

Artificial neural networks are used to train the feature extracted images for characterizing them as systole or diastole. Depending on the weight provided and the training values (input sample), the training pattern changes. Back-propagation algorithm is used in this work as the target is met with training and error control action.

9.7 Additive Gaussian Noise

Image acquisition threats are mostly by the uncertain production of Gaussian noise. To epitomize, the reasons are plenty—poor illumination, an interrupted signal transmission, or heat dissipation can cause such noises. Incorporating a spatial filter can nullify this to a great extent. However, an image smoothening assures the blurred image edges as it blocks high-frequency signals.

9.8 Input Image

The 20 image samples of normal and cardiac lesions were acquired using the echo machine of Philips. They were obtained in DICOM format. For our purpose, it was converted to png format. Image preprocessing step includes resizing the image to 256 × 256 images, gray scale conversion, and filtering. Image resizing is done using imresize which computes the image aspects to preserve the aspect ratio.

For image enhancement, imadjust is used which increases the contrast of the output image. The logic is of course that noise and other small variations are unlikely to result in a strong edge. Thus, strong edges will only be due to true edges in the original image. The weak edges can either be due to true edges or noise/color variations.

The latter type will probably be distributed independently of edges on the entire image, and thus only a small amount will be located adjacent to strong edges. Weak edges due to true edges are much more likely to be connected directly to strong edges.

9.9 Artificial Neural Network Training

Artificial neural networks are relatively crude electronic models based on the neural structure of the brain. The brain basically learns from experience. It is natural proof that some problems that are beyond the scope of current computers are indeed solvable by small energy efficient packages. This brain modeling also promises a less technical way to develop machine solutions. This new approach to computing also provides a more graceful degradation during system overload than its more traditional counterparts. "logsig" is a transfer function. Transfer functions calculate a layer's output from its net input. purelin is a neural transfer function. Transfer functions calculate a layer's output from its net input. trainrp is a network training function that updates weight and bias values according to the resilient back-propagation algorithm (Rprop). "trainrp" can train any network as long axis weight, net input, and transfer functions have derivative functions. Figure 9.2 shows 2D ECG image extraction and its classification using ANN.

Various noise quality parameters were calculated for Gaussian noise scenario and multiplicative Rayleigh noise scenario. A better denoising is characterized by higher values of SNR and PSNR parameters and lower values of the MAE and MSE parameters. The aforementioned characteristics were satisfied in Gaussian noise scenario.

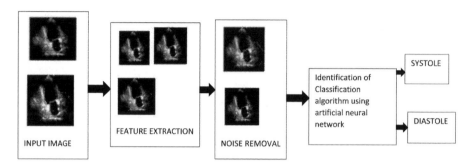

FIGURE 9.2
2D ECG image extraction and its classification using ANN.

9.10 Conclusion

The study aims at discerning systole and diastole from 2D images. The major and minor axis height and width are calculated. Analysis shows that the preprocessing steps and edge detection steps are well applicable for apical two chamber long-axis, 0° view (LAX0) of 2D echocardiographic images. The region segmentation highlights the feasible myocardium wall region in the gray scale image. Various noise quality parameters were calculated for Gaussian noise scenario and multiplicative Rayleigh noise scenario. A better denoising is characterized by higher values of SNR and PSNR parameters and lower values of the MAE and MSE parameters. The aforementioned characteristics were obtained in Gaussian noise scenario. The back-propagation network works well for the classification as it updates the weights in each step to attain the best result.

9.11 Case Study 2: Detection of Lung Cancer from CT Thoracic Images Using Hybrid Soft Computing Techniques

The mortality rate due to cancer is irrespective of the gender where lung cancer is the leading cause of death. Even though different treatments are available, none shows an accurate therapy for this deadly disease. In fact, the treatment options vary with the type of lung cancer—small cell lung cancer (SCLC) and nonsmall cell lung cancer (NSCLC). A minor category of lung cancer belongs to SCLC which is aggressive and grows and spreads faster to other organs. However, this accounts for only 15% of the total victims; to a great extent this case is associated with smoking habits. Less than 6% of patients diagnosed with lung cancer have never smoked. The second category constitutes for abnormal tumor cell formation in the lungs and spreads to a wide area. The brain child of image processing techniques can be used to determine and detect these lesions in lungs. An analysis of CT image of lungs is taken for study. An automatic approach of detection from CT thoracic images is designed that covers primary lung tumors and subsequent lymph nodes.

Image processing algorithms give an image in the prescribed form after series of processing techniques. Though it gives a good visual inspection, classification of certain attributes is always troublesome. This task can be accomplished using cognitive techniques, while the user effort is minimized. A neural network based image processing is developed in this work that employs method of multistage discriminative model to differentiate normal and abnormal images. The training of network is sophisticated many a time if the data are not accurately given. The algorithm is developed such that the neural network is trained using different set of CT lung images.

9.12 Overview of the Work

The visual inspection of lung tumor through medical imaging technique is a tedious and error prone task. Most of the times the radiologist can misunderstand abnormal lymph nodes as tumor because of same appearance. In this project, we introduced a system which automatically detects tumor and differentiates tumor from abnormal lymph nodes. Basic image processing techniques like preprocessing, feature extraction, morphological image processing operations like opening and closing are done in CT image in order to reduce noise and to find the orientation of the image. Neural network concepts are used for testing and training the image through Gray Level Co Matrix (GLCM) features and wavelet features. These features of normal source image are subtracted with the abnormal source image to find the exact range and orientation of tumor cells in the lungs. For these processes, we are developing computer-aided diagnosis software (PYTHON) that detects tumor in the lungs. The tumor detection process involves the following steps:

Preprocessing

Morphological operations

Feature extraction

Neural networks classification

Figure 9.3 shows the overview of the proposed system.

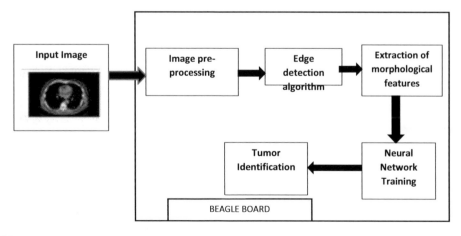

FIGURE 9.3
Overview of the proposed system.

9.13 Preprocessing

The image orientation based processing is carried out which eliminates noise and improves the quality of image. Like in the case discussed before, the size of the input sample is large and may produce inaccurate results. Hence, the search area and sample size should be reduced before processing the original image. Accuracy of image segmentation improves the quality of image under study.

A median filter equipped with noise removal techniques together with removal of surplus parts make the image ready for diagnosis.

9.14 Morphological Operations

The biological image processing includes diagnostics and morphological image analysis. The morphology or shape of some parts/organs analysis is referred to as morphological image processing. These include dilation, erosion, opening, and closing. A threshold image close to the target area is identified and the small gaps and holes are filled in this image. A binary lung mask ensures that bigger sized blocks are kept undisturbed, and other eight-connected neighboring pixels are set to zero. The boundary pixels can be segmented if all the neighboring pixels' value is zero, then that boundary can be extracted. The mask superimposed on lung CT image gets the final segmented image that is required for diagnosis. A morphological operation acts as a structure in developing an output image of equivalent size.

9.15 Feature Extraction

Feature extraction refers to the process of locating the pixels that have distinguished features or characteristics. The homogenous characteristics as image intensity and range are considered as local image properties. Typically, characteristics that are homogenous are the local image properties. The extracted features account for the identification of dissimilarities and similarities, pixel range classification, comparison of identify, and localization of anatomical structures.

9.16 Neural Network Classification

Neural networks are widely used in medical field for diagnosis of complex diseases like cancer. The major outbreak of neural network is the detection of lung cancer. The most noted challenge in the diagnosis and detection of such cases is the practical limitations in acquiring and retrieving medical data. The medical ethics requires confidentiality of each case. The legal samples acquired are screened and segregated properly to select the best classifier to

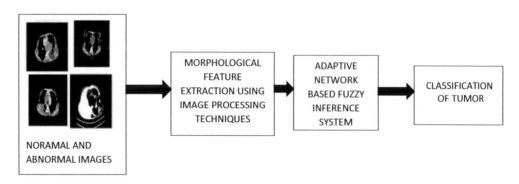

NORAMAL AND
ABNORMAL IMAGES

MORPHOLOGICAL
FEATURE
EXTRACTION USING
IMAGE PROCESSING
TECHNIQUES

ADAPTIVE
NETWORK
BASED FUZZY
INFERENCE
SYSTEM

CLASSIFICATION
OF TUMOR

FIGURE 9.4
Detection of Lung Tumour with ANFIS.

analyze the database. In certain cases, the affected people will be less in number but the severity will be higher and vice versa. There will be an imbalance when the patient number is large and screening population is limited. It is obvious that artificial neural networks are widely exploited for lung cancer diagnosis. The most challenging process in medical field is collection of data, which has some practical limitations. By replacing the rule base classifiers in conventional classifier systems, neural networks are implemented. A population of classifiers or rules is the composition of classifier system. There is a "strength" associated with each classifier. The energy or power of each classifier is expressed through the evolution process, which varies in state considering smaller size. The performance can be improved by increasing the size of the training. Figure 9.4 shows the detection of lung tumor with ANFIS.

The detection of normal and abnormal cells in the lungs is obtained from CT thoracic images which are trained using neural networks. The classification accuracy increases with the increase in the number of trained data and hence normal and diseased images of more numbers are taken to train the classifier. The training gives a fair differentiation of the images as tumors or abnormal nodes based on the loaded data. A hybrid system is designed using the integrated fuzzy rules using neural network training for better accuracy. The training data include wide range of cell images like normal cells, abnormal lung cell, and diseased cells. The images function as a database for training and differentiation for detection of lymph nodes, tumor and lung tumor classification.

9.16 Results and Discussions

The detection and classification of lung tumors and further testing was done with the aid of PYTHON software, and images were retrieved from database. The processing includes initial detection and classification of cancer images, and comparison with the database from the given input features for the trained images. The classification result is obtained through the basic image processing techniques in order to find exact range and orientation of the tumor cells. The difference between features of normal images and abnormal images shows the tumor region, which was further processed through PYTHON coding to get the exact output.

9.17 Conclusion

The detection of tumor from CT images was done and classified using PYTHON software. The training data were increased for accuracy resulting in a decrease in error rate and thereby an increase in computational speed. The results were compared with other methods and the accuracy was obtained. The comparison proves that the method gives a better outcome when other methods like support vector machine and conditional random fields are considered. This throws an insight to the detection and diagnosis of lung tumor and thereafter an effective surgery for the same. The reduction in negative errors in images helps in distinguishing the abnormal lymph nodes as lung tumor. Further classification can be done for other imaging modalities like MRI and PET for lung cancer detection and also tumors in brain, abdomen, and uterus which are predominant these days. The 3D thoracic images can also be considered as an extension of this work.

References

[1] Yang Song, Weidong Cai, Jinman Kim, and David Dagan Feng, "A Multistage Discriminative Model for Tumor and Lymph Node Detection in Thoracic Images," IEEE Transactions on Medical Imaging, Vol. 31, no. 5, May 2012.

[2] Hui Cui, Xiuying Wang, and Dagan Feng, "Automated Localization and Segmentation of Lung Tumor from PET-CT Thorax Volumes Based on Image Feature Analysis," 34th Annual International Conference of the IEEE EMBS, San Diego, CA, 28 August–1 September, 2012.

[3] Cherry Ballangan, Xiuying Wang, Michael Fulham, Stefan Eberl, Yong Yin, and Dagan Feng, "Automated Delineation of Lung Tumors in PET Images Based on Monotonicity and a Tumor-Customized Criterion," IEEE Transactions on Information Technology in Biomedicine, Vol. 15, no. 5, September 2011.

[4] Ivana Iğgum, Marius Staring, Annemarieke Rutten, Mathias Prokop, Max A. Viergever, and Bram van Ginneken, "Multi-Atlas-Based Segmentation with Local Decision Fusion—Application to Cardiac and Aortic Segmentation in CT Scans," IEEE Transactions on Medical Imaging, Vol. 28, no. 7, July 2009.

[5] Eva M. van Rikxoort, Bartjan de Hoop, Saskia van de Vorst, Mathias Prokop, and Bram van Ginneken, "Member, IEEE Automatic Segmentation of CT Pulmonary Segments from Volumetric Chest Scans," IEEE Transactions on Medical Imaging, Vol. 28, no. 4, April 2009.

[6] Iyad Jafar, Hao Ying, F. Shields Anthony, and Otto Muzik, "Computerized Detection of Lung Tumors in PET/CT Images," Proceedings of the 28th IEEE EMBS Annual International Conference, New York, August 30–September 3, 2006.

[7] K. Begg Rezaul, Marimuthu Palaniswami, and Brendan Owen, "Support Vector Machines for Automated Gait Classification," IEEE Transactions on Biomedical Engineering, Vol. 52, no. 5, May 2005.

[8] J. Kuhnigk, V. Dicken, L. Bornemann, A. Bakai, D. Wormanns, S. Krass, and H. Peitgen, "Morphological Segmentation and Partial Volume Analysis for Volumetry of Solid Pulmonary Lesions in Thoracic CT Scans," IEEE Transactions on Medical Imaging, Vol. 25, no. 4, pp. 417–434, April 2006.

[9] Y. Boykov, O. Veksler, and R. Zabih, "Efficient Approximate Energy Minimization via Graph Cuts," IEEE Transactions on Pattern Analysis and Machine Intelligence, Vol. 20, no. 12, pp. 1222–1239, December 2001.

[10] W. Wever, S. Stroobants, J. Coolen, and J. Verschakelen, "Integrated PET/CT in the Staging of No Small Cell Lung Cancer: Technical Aspects and Clinical Integration," European Respiratory Journal, Vol. 33, pp. 201–212, 2009.

[11] D. Bibicu and L. Moraru, "Cardiac Cycle Phases Estimation in 2D Echocardiographic Images Using Artificial Neural Network," IEEE Transactions on Biomedical Engineering, 2012.

[12] Sujata N. Patil, Uday V. Wali, and Mk Swamy, "Application of Vessel Enhancement Filtering for Automated Classification of Human In-Vitro Fertilized (IVF) Images," International Conference on Electrical Electronics, Communication, Computer and Optimisation Techniques (ICEECCOT), 2016.

[13] Asha Vincet and Reshmi Mariyam Reji Jacob, "Determination of Atrial Diastole and Systole from 2-D Echocardiographic Images Using Artificial Neural Network," International Journal for Research in Applied Science and Engineering Technology, Vol. 2, no. 4, 2014.

[14] Ismo Kinnunen and Anssimakyen, "Image Based Size Distribution Measurement of Gravel Particles," IEEE International Instrumentation and Measurement Technology Conference, 2011.

[15] V. Gurunathan, S. Bharathi, and R. Sudhakar, "Image Enhancement Techniques for Palm Veinimages," 2015 International Conference on Advanced Computing and Communication Systems, 2015.

10

Alzheimer's Disease Classification Using Machine Learning Algorithms

S. Naganandhini and P. Shanmugavadivu

The Gandhigram Rural Institute (Deemed University), Dindigul, Tamil Nadu, India

A. Asaithambi

University of North Florida, Jacksonville, FL, USA

M. Mohammed Mansoor Roomi

Thiagarajar College of Engineering, Madurai, Tamil Nadu, India

10.1 Introduction

Alzheimer's disease (AD) is a severe brain disease that adversely impacts the ability to think and may even lead to death in its final stage. Therefore, early diagnosis of AD is critical for proper treatment. Machine learning (ML) is a branch of artificial intelligence that employs a variety of probabilistic and optimization techniques to detect post-concussion syndrome (PCs) from huge and multipart datasets. This has motivated many researchers to focus on using machine learning for early diagnosis of Alzheimer's disease. The performance of machine-learning-based methods for AD classification depends on various factors such as training data, preprocessing, feature selection, and classifiers. In this chapter, machine learning is proposed as a model that includes preprocessing, attributes selection, and classification through the association of rule mining concepts, and as a model that provides a solution for the diagnosis of AD.

According to the U.S. National Institutes of Health–National Institute on Aging (NIH-NIA), "Alzheimer's disease (AD) is a progressive degeneration of brain functions leading to cognitive and physical dysfunctions, especially among the aged." Currently, AD assumes third place among common disabilities across the world, next to cardiovascular diseases and cancer. It is also recorded as the sixth major cause of mortality in the United States[1] The development of neurodegenerative diseases cannot be detected directly by the visual analysis of radiologists. Neuroimaging is a vital tool in the early diagnosis of neurodegenerative diseases by deriving objective patterns and structural relations from magnetic resonance (MR) images. As observed by Reuda et al.,[2] abnormality detection in MR brain images is a challenging task. AD is the most common type of dementia and among the most common disabilities across the world. Early detection of AD is needed to provide the necessary treatments to the patients at the right time. Structural magnetic resonance imaging (sMRI) is a powerful diagnostic tool which gives images with high resolution and high tissue contrast. Generally, AD can be characterized into three stages: mild, moderate, and severe. This chapter presents four methods for the classification of AD using a dataset with T1_weighted MRI images from Open Access Series of Imaging Studies (OASIS).

10.2 Review of the Literature

Automated brain disorder diagnosis with MR images is becoming increasingly important in the medical field. Several works related to AD classification are presented next.

Tohka et al.[3] comparatively analyzed the performance of support vector machine (SVM) and features-based classification algorithms that detect the level of dementia using functional MRI (fMRI) datasets. This work reported on the classification of AD and mild cognitive impairment (MCI) against normal control (NC), by using split-half resampling from the Alzheimer's Disease Neuroimaging (ADNI) database. The study focused on the impact of feature selection with and without filter as well as embedded features against SVM. The study concluded that the selection of sample size and the number of features substantially influence the accuracy of classification.

Chaplot et al.[4] used the MR brain images transformed into wavelets as input to an SVM and a self-organizing map (SOM) to classify those MR brain images as normal or abnormal. The conclusion of this study was that SOM outperformed SVM in yielding better classification accuracy.

Herrera et al.[5] devised a tool for prognosis and classification of dementia. In addition to classifying the MRI as normal or abnormal, this study also classified the images as normal, MCI, or AD. It is observed that the precision of classification is influenced by feature extraction mechanism, dimensionality reduction, and dataset partition for training and testing.

Li et al.[6] proposed a modification in the classical deep learning approach for AD diagnosis using MRI and positron emission tomography (PET) images. The authors attempted to explore the influence of dropout as a solution to handle overfitting caused by weigh coadaptation. They proposed a new deep learning framework infused with stability selection, adaptive learning, and multitask learning to classify the input images into AD MCI. This study confirmed that incorporating dropout in deep learning improves the accuracy of classification.

Ye et al.[7] explain the causes and consequence of AD, and report on the new approaches of AD detection on neuroimaging using multisource data fusion and multisource biomarker selection. This study uses techniques based on sparse inverse covariance estimation (SICE) to discover network structure connectivity and its strength.

Lama et al.[8] summarize the advantages of machine learning techniques to detect AD in neuroimaging namely, PET, MRI, and fluorodeoxy glucose (FDG PET) scan. The authors indicate that the observations on functional brain connectivity ideally help in the identification of AD, diagnosis of AD, and classification of images as MCI and healthy control (HC). The principles used include SVM, import vector machine (IVM), and regularized extreme learning machine (RELM). This work also includes the development of another classification mechanism exclusively for complex datasets, using the ADNI database for experimental analysis.

Torabi et al.[9] developed an AD detection technique using texture-based feature extraction. In the texture analysis system (TAS) proposed, feature vector of the input neuroimages is constructed and used as input to the feed-forward neural network. This work uses the gray-level co-occurrence matrix (GLCM) and principal component analysis (PCA). With a 60:40 ratio of training and test data, this work achieved an accuracy of 95%.

Torabi et al.[10] present an AD detection method for neuroimages. In this work, the spatial and frequency features extracted from the input images are optimized using feature reduction. The optimized features are then fed to a multilayered perception based neural network. The accuracy recorded for training and testing was 100% and 79%, respectively.

Huang et al.[11] devised a new technique for AD classification, which is a hybrid of voxel-based morphometry (VBM), PCA, and feed-forward artificial neural network (ANN) with back-propagation for feature extraction and classification. The accuracy of classification was consistent across the trials in AD prognosis.

El-Sayed et al.[12] presented a tri-phased AD classifier for MRIs, by combining feature extraction, dimensionality reduction, and AD classification. They used discrete wavelet transformation (DWT) for feature extraction; PCA for feature reduction; and feed forward back-propagation artificial neural network (FP-ANN) and k-nearest neighbor (k-NN) for classification. The accuracy of classification by these methods was 97% or better.

Sampath et al.[13] developed a novel method for image segmentation and AD detection from fMRI. This work used SOM network for feature extraction and segmentation; and the adaptive neuro-fuzzy inference system (ANFIS) was used for classification. This combinatorial AD detection approach gave appreciable results.

Freund et al.[14] reported on the learning aspect of a generic decision-theoretic approach and its applications. The authors introduced a multiplicative weight update process in boosting, and analyzed the behavior of this modified method in the light of learning algorithm and functions.

10.3 Description of Classifiers

Classifiers are models that allocate appropriate class labels (from a set of known class labels) to test samples represented as feature vectors.

10.3.1 Naïve Bayes Classifiers

Naïve Bayes (NB) is an efficient technique for building classifiers. The technique uses groups of algorithms that are based on the principle that the feature value is independent of the given class variable. The Bayes classifier model is a supervised learning model and has many practical applications due to its reliance on maximum likelihood based parameter estimation. This model can work even when one does not assume Bayesian probability or does not use any Bayesian methods. In spite of these assumptions regarding parameters, Naïve Bayes classifiers are able to handle many complex real-world situations quite well. Bayes classification can provide better performance even for small training samples. On the other hand, while the efficacy of Naïve Bayes classifiers was proved in 2004, Bayes classification was outperformed in 2006 by other classifiers such as boosted trees or random forests.

10.3.2 Gaussian Naïve Bayes

It is customary to assume that continuous data associated with each class obey Gaussian distribution. Let x be a continuous attribute in the training data. These data may be first segmented by class, and then the mean and variance of x may be computed in each class. Let μ_k be the mean of the values in x associated with class C_k and let σ_k^2 be the variance of the values in x associated with class C_k. Suppose we have collected some observation value v. Then the probability distribution $p(x = v|C_k)$ of v given a class C_k can be computed using

$$P(x = v|C_k) = \frac{1}{\sqrt{2\pi}\sigma_k} e^{-\frac{1}{2}\left(\frac{v - \mu_k}{\sigma_k}\right)^2}$$ (10.1)

10.3.3 Multilayer Perceptrons (MLP)

A perceptron is an algorithm that classifies given input by separating into two categories with a linear function, and thus called a linear classifier. The input is typically a feature vector x multiplied by weights w and added to a bias b, mathematically expressed as $w^T x + b$. Additionally, a perceptron sometimes passes $w^T x + b$ through a nonlinear activation function y and deals with it in the form $y(w^T x + b)$.

10.3.3.1 Activation Function

In MLPs, some neurons use nonlinear activation functions developed to model how frequently the biological neurons are fired. The two commonly used activation functions are both sigmoid, and are described by

$$y(v_i) = tanh(v_i)$$ (10.2)

$$y(v_i) = \frac{1}{1 + e^{-v_i}}$$ (10.3)

The hyperbolic tangent ranges in values from −1 to 1. The second function, often referred to as the logistic function, has a shape similar to the hyperbolic tangent function, but ranges in value from 0 to 1. Here $y(v_i)$ is the output of the ith node (neuron) and v_i is the weighted sum of the input connections. Other activation functions, such as the rectifier and soft plus functions, and radial basis functions have been proposed.

The MLP consists of an input layer, and possibly one or more hidden layers of nonlinearly activating nodes or sigmoid nodes. MLPs are used to handle supervised learning problems. A given set of input-output pairs is used for training so that the correlations and/or dependencies between the inputs and outputs can be modeled. During such modeling, the weights may need to be adjusted while training in order to minimize error. The weights are typically adjusted by an iterative process. Small changes in the weights to get the desired values are accomplished using a process called training the net and the training set (learning rule). For more information on MLPs, the reader is referred to Iddamalgoda et al.[15]

10.3.4 Random Forest Classifier

The random forest classifier (RFC) performs regression tasks as well as classification tasks using a supervised learning approach. It also undertakes dimensionality reduction if needed, and can handle missing values, outliers, and other important processes in data exploration. It may be considered as a group learning method, where a group of weak models aggregate to form an efficient model.[16]

A random forest includes multiple trees (Figure 10.1) whereas the CART model involves a single tree. The trees in the forest participate in the classification by casting a vote for a class label in order to classify a new object. The class with a majority vote (over all the trees in the forest) is selected as the class label for the object being classified. In the case of regression, the forest takes the average value of the outputs provided by different trees as the needed result.

10.3.4.1 Random Forest Hyperparameters

The hyperparameters in the random forest algorithm are used to accelerate the predictive power of the model and increase the model's speed.

A. **Increasing the Predictive Power:** The hyperparameter $n\_estimators$ denotes the number of trees included in the algorithm before computing the majority voting or taking average values of the prediction labels. Generally, the time taken by the algorithm is directly proportional to the number of trees used to build the forest. While this approach helps to increase accuracy, it slows down the computation. The next hyperparameter, namely, max feature represents the maximum limit on the

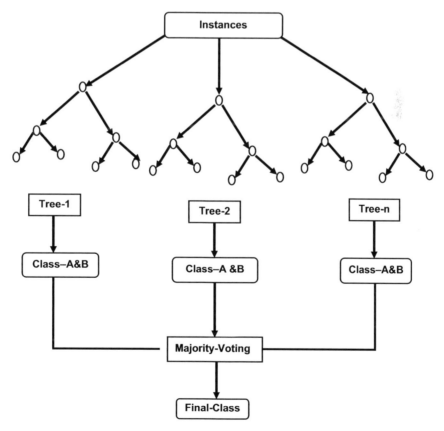

FIGURE 10.1
Random Forest Classifier Model.

number of features allowed to try for random forest classification in an individual tree. The last hyperparameter which is related to speed is min sample leaf. This parameter depicts the minimum number of leaves required to split an internal node.

B. **Increasing the Model's Speed:** The *n_jobs* hyperparameter determines the number of processors allowed. When this parameter has a value of 1, it means only one processor can be used; a value of –1 means there is no limit on the number of processors. The hyperparameter random state determines the replicability of the output of the model. The model will provide the same result when it has a definite value of random state and if the same hyperparameters and training data are used. The cross validation for the random forest model is determined by the oob_score hyperparameter. In this cross validation, one-third of the data is retained to test the model and not used for training. These retained samples are called out of the bag samples.

10.3.5 Algorithm for Random Forest

In the first stage, the random forest algorithm randomly selects k features out of entire set of m features. In the second stage, the randomly selected k features are used to find the root node using the best split strategy. In the third stage, the daughter nodes are calculated using the same best split approach. The fourth stage is the repetition of the first three stages until the formation of the tree on a root node with a target number of leaf nodes is completed. In the last stage, the first four stages are repeated so that n trees are created randomly, thus forming the random forest.

---

1. From a total m features, k features are randomly selected (k << m).
2. The root node is calculated from the k features selected using the best split strategy.
3. Split the nodes into daughter nodes using the best split.
4. Steps 1 through 3 are repeated until nodes have been reached.
5. Steps 1 through 4 are repeated n times to create n trees.

10.3.6 Gradient Booster Classifier (GBC)

Two buzzwords, Bagging and Boosting are frequently encountered while working with boosting algorithms. The term Bagging refers to building learning algorithms on random samples of data and taking simple means to determine bagging probabilities. The term Boosting, on the other hand, refers to a similar process, but sample selection is carried out more intelligently. This enables us to give increasing weights to hard-to-classify observations.[17]

The gradient booster classifier does boosting as opposed to bagging (see Figure 10.2). This boosting technique employs the logic in which subsequent predictors learn from the mistakes of previous predictors. Therefore, the observations have an unequal probability of appearing in subsequent models and the ones with the highest errors appear the most. It should be noted that the observations are not chosen based on the bootstrap process, but based on the errors. The predictors can be chosen from a range of models like decision trees, regressors, classifiers, etc. Because new predictors are learning from mistakes committed by previous predictors, it takes less time/iterations to

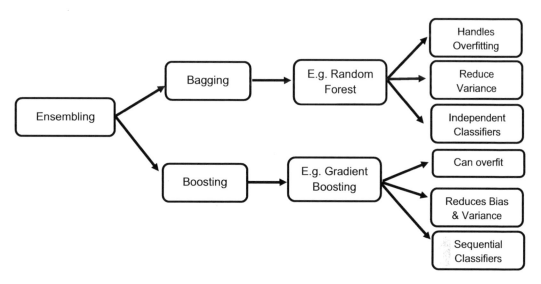

FIGURE 10.2
Gradient Booster Classifier.

reach close to actual predictions. However, the stopping criteria should be chosen carefully or the process could lead to overfitting on training data.

10.3.7 Gaussian Process Classifier (GPC)

Joint probability distribution and the Bayes rule form the basis for probabilistic classification. The Bayes theorem relates the joint probability and conditional probabilities of two stochastic events. Let x be the feature vector and y be the class label. Then, the Bayes theorem can be written as

$$p(y|x) = \frac{p(x,y)}{p(x)} = \frac{p(y) \cdot p(x|y)}{p(x)} \tag{10.4}$$

where $p(x|y)$ and $p(y|x)$ are conditional probabilities, and $p(y)$ are prior probabilities of events x and y. Generative models find the posterior probability $p(y|x)$ for each class from the conditional probability of the class and the prior probabilities. The generative model can be categorized into two types, namely, parametric and nonparametric.

The method of Gaussian process classifier is a nonparametric classification method that defines the posterior probabilities over a latent function $p(f)$ where the function $p(f)$ is assumed to be normal or Gaussian distribution. This assumption guarantees the smoothness properties for the underlying probability density function. It is subjected to a logistic likelihood function Ω like sigmoid function for classification. Let us consider a binary hypothesis deciding an entity to either $C_1(y_i = 0)$ or $C_2(y_i = 1)$ and x_i the feature vector of ith training sample with its class as C_j, where i varies from 1 to n; j is either 1 or 2; and n is the number of training samples. This Gaussian process is characterized by the mean function $\mu(x)$ and the covariance function $k(x, x')$.

$$k(x, x') = \sigma_j^2 \exp\left[-\frac{1}{2}\frac{(x - x')^2}{i^2}\right] \tag{10.5}$$

where σ_j^2 represents the maximum allowable covariance and l the length parameter. If $k(x, x')$ is high, then it means that $f(x)$ and $f(x')$ are perfectly correlated. The Gaussian process classifier learns the hyperparameters from the training samples, computes the marginal likelihood, and does posterior inference. Let x_{n+1} be a test feature vector with its corresponding latent variable f_{n+1}. The objective is to determine the class label y_{n+1}. The posterior probability for class C_1 is

$$P(C_1|x) = P(y_i = 0|f_i) = \Omega(f_i(x)) \tag{10.6}$$

First, the distribution corresponding to the test feature vector is computed as

$$p(f_{n+1}|X, y, x_{n+1}) = \int p(f_{n+1}|X, y, x_{n+1}, f)p(f|X, y)df \tag{10.7}$$

Later, the posterior inference is made by imposing this distribution over the latent variable f_{n+1} as

$$p(y_{n+1}|f_{n+1}) = \int \Omega(f_{n+1})p(f_{n+1}|X, y, x_{n+1}, f)df_{n+1} \tag{10.8}$$

Two analytic approximations using non-Gaussian joint posterior with a Gaussian are: Laplace approximation and expectation propagation. These can also be employed with Gaussian process for classification.

10.3.8 Adaptive Boosting Classifier (AdaBoost)

An AdaBoost[14] classifier first fits a classifier on the original dataset. Then, it also fits additional copies of the classifier on the same dataset. In this second pass, AdaBoost adjusts the weights of those instances that were incorrectly classifier previously so that subsequent classifiers focus more on difficult cases[18] (see Figure 10.3).

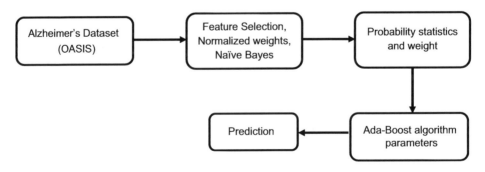

FIGURE 10.3
Adaptive Boosting Classifier Model.

AdaBoost Classifier Hyperparameters:

1. **base estimator: object, optional (default=None)**
   The boosted ensemble is constructed from this estimator.

2. **n estimators: integer, optional (default=50)**
   The limit on the number of estimators before boosting is terminated. If perfect fit were to occur sooner, the learning procedure may be stopped early.

3. **Learning rate: float, optional (default=1)**
   Learning rate shrinks the contribution of each classifier by learning rate. There is a trade-off between learning rate and n estimators.

4. **algorithm: "SAMME", "SAMME.R", optional (default="SAMME.R")**
   This parameter allows the choice between the discrete versus real boosting algorithm. When this parameter has a value "SAMME.R," the real boosting algorithm is used, and the base estimator parameter is expected to support the calculation of class probabilities. When "SAMME" is chosen, then the discrete boosting algorithm is used. The real algorithm typically converges faster than the discrete algorithm, and yields a lower test error while requiring fewer boosting iterations.

5. **random state: int, RandomState instance or None, optional (default=None)**
   This parameter indicates information used for random number generation. When *int* is chosen for random state, then it represents the seed that needs to be used by the random number generator. On the other hand, when RandomState is chosen for random state, it specifies the random number generator to be used.

10.4 Results and Discussion

This section presents the results obtained when the different classifiers are used on a specific dataset. After presenting information pertaining to the dataset used, we present our results on classification accuracy, precision, recall, and F1-Score. A discussion of results is also included in this section.

10.4.1 Dataset Description

The dataset we used consists of an MRI collection from 416 subjects aged 18 to 96. The MRIs in the collection are cross-sectional images in young, middle-aged, as well as non-demented and demented older adults. The subjects involved come from both genders, and are all right-handed. Of the 416 subjects, 100 subjects over the age of 60 have been clinically diagnosed with very mild to moderate AD. The dataset also consists of longitudinal MRI data from 150 non-demented and demented older adults aged 60 to 96. Each of these subjects was scanned on two or more visits with at least one year gap for 373 imaging sessions. Of these, 72 subjects were categorized as non-demented throughout the study; 64 were categorized as demented at the first visit and remained so for subsequent scan sessions, including 51 individuals with mild-to-moderate Alzheimer's disease. Finally, 14 subjects from this collection were considered as non-demented at their first visit and were characterized later as demented at a later visit. Table 10.1 provides information on the demographics and Table 10.2 provides clinical information.

TABLE 10.1

Demographics Information

Descriptor	Explanation
M/F	Gender
Hand	Handedness (actually all subjects were right-handed so this item can be omitted)
Age	Age in years
EDUC	Years of education
SES	Socioeconomic status as assessed by the Hollingshead Index of Social Position and classified into categories from 1 (highest status) to 5 (lowest status)

TABLE 10.2

Clinical Information

Descriptor	Explanation
MMSE	Mini-Mental State Examination score (range is from 0 = worst to 30 = best)
CDR	Clinical Dementia Rating (0 = no dementia, 0.5 = very mild AD, 1 = mild AD, 2 = moderate AD)

10.4.2 Confusion Matrix

A confusion matrix summarizes the predicted results on a classification problem. The number of right and wrong predictions is summarized with count values with respect to each class. The confusion matrix depicts how the classification model is confused in making predictions. Calculating the confusion matrix: In a two-class problem, we wish to differentiate observations that belong to one class from those that belong to a second class. For our situation, this may correspond to the presence of the disease versus its absence, which we call as event (occurrence) and no event (absence), respectively. The confusion matrix will have two rows and two columns. The rows represent even or no event. In other words, we may say that the "occurrence" row is regarded as positive or Class 1, and the "absence" row as negative or Class 2. Similarly, the columns, corresponding to the classification results, are regarded true and false;[19] and Table 10.3 summarizes this information.

Similarly, Table 10.4 shows a sample confusion matrix with numerical values and illustrates how certain metrics can be calculated from the confusion matrix.

TABLE 10.3

Elements of Confusion Matrix

Descriptor	Explanation
True positives (TP)	Classifier predicted the test samples as diseased when they have the disease
True negatives (TN)	Classifier predicted no disease when they don't have the disease
False positives (FP)	Classifier predicted the test samples as diseased, but they don't have the disease
False negatives (FN)	Classifier predicted no disease, but they have the disease

TABLE 10.4

Metrics from Confusion Matrix

Metric	Explanation
Accuracy	(TP + TN)
Misclassification Rate	(FP + FN) = Total
True Positive Rate or Recall	TP = Actual Yes
False Positive Rate	FP = Actual No = 10 = 60
True Negative Rate	TN = Actual No = 50 = 60 0:83

10.4.3 The F1-Score

We conclude this section with a brief description of the F1-Score metric. This score, also referred to as the F-Score or F-Measure, is an evaluation metric for assessing the performance of a classifier. The F1-Score is calculated using both the precision p and the recall r achieved by the classifier, where p is the number of true positive results among all positive results decided by the classifier, and r is the number of true positive results among all relevant samples[20] showing the computation of p and r. The F1-Score is the harmonic mean of the precision and recall. The higher the values of p and r, the higher the F1-Score.

10.4.4 Experimental Results

The different classifiers were tested using the dataset described, and the classification precision, accuracy, recall, and F1-Score were determined for each classifier. The results presented here will provide a comparison of performance of the different classifiers with respect to each metric. The metrics are calculated as the percentage of number of correctly classified samples divided by the total number of samples.

Figure 10.4 and Table 10.5 show a comparison of the classification precision obtained by various classifiers.

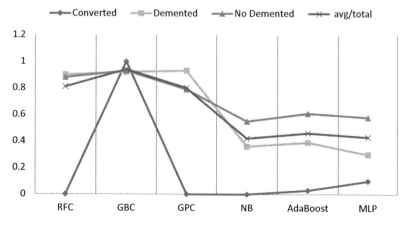

FIGURE 10.4
Performance Analysis of Classifiers Using Precision.

TABLE 10.5

Precision Values of Classifiers

Group	RFC	GBC	GPC	NB	AdaBooster	MLP
Converted	0.00	0.50	0.00	0.00	0.08	0.11
Demented	1.00	1.00	0.81	0.36	0.45	0.32
Non-demented	0.98	0.98	0.97	0.60	0.33	0.54
Avg/total	0.89	0.93	0.84	0.45	0.34	0.41

Figure 10.5 and Table 10.6 show a comparison of the classification recall obtained by various classifiers.

Figure 10.6 and Table 10.7 show a comparison of the classification F1-Score obtained by various classifiers.

Figure 10.7 and Table 10.8 show a comparison of the classification accuracy obtained by various classifiers.

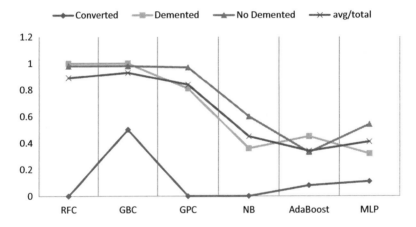

FIGURE 10.5
Performance Analysis of Classifiers Using Recall.

TABLE 10.6

Recall Values of Classification

Group	RFC	GBC	GPC	NB	AdaBooster	MLP
Converted	0.00	1.00	0.00	0.00	0.03	0.10
Demented	0.90	0.92	0.93	0.36	0.39	0.30
Non-demented	0.88	0.93	0.79	0.55	0.61	0.58
Avg/total	0.81	0.94	0.80	0.42	0.46	0.43

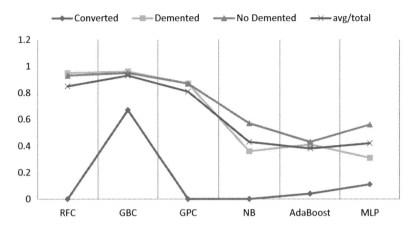

FIGURE 10.6
Performance Analysis of Classifiers Using F1-Score.

TABLE 10.7

F1-Score Values of Classification

Group	RFC	GBC	GPC	NB	AdaBooster	MLP
Converted	0.00	0.67	0.00	0.00	0.04	0.11
Demented	0.95	0.96	0.87	0.36	0.41	0.31
Non-demented	0.93	0.95	0.87	0.57	0.43	0.56
Avg/total	0.85	0.93	0.81	0.43	0.38	0.42

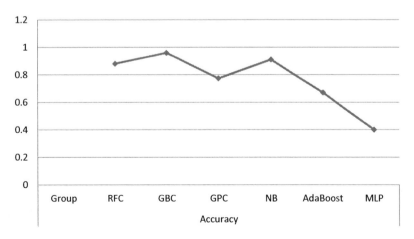

FIGURE 10.7
Performance Analysis of Classifiers Using Accuracy.

TABLE 10.8

Accuracy Values of Classification

Group	RFC	GBC	GPC	NB	AdaBooster	MLP
Accuracy (%)	88	96	77.33	91.07	66.96	40

TABLE 10.9

Performance Summary of the Classifiers

Metric Classifier	Precision			Recall			F1-Score		
	C	D	ND	C	D	ND	C	D	ND
RFC	S	E	H	S	E	E	S	E	E
GBC	E	E	E	M	E	E	M	E	E
GPC	S	E	H	S	H	E	S	H	H
NB	S	L	M	S	L	M	S	L	M
AdaBoost	S	L	M	S	L	L	S	L	L
MLP	P	L	M	P	L	M	S	L	M

C: Converted; D: Demented; ND: Non-Demented
Excellent (E): 90–100; High (H): 70–89; Moderate (M): 50–69;
Low (L): 30–49; Poor (P) 10–29; Scant (S): 0–9

For a side-by-side comparison of the performance of the different classifiers, we use six quality indicators—Scant, Poor, Low, Moderate, High, and Excellent. We say a classifier exhibits a certain level of performance in a particular metric if it achieves a percentage value as specified next for that metric.

S: Scant (S) performance if value achieved is 0–9%;
P: Poor performance if value achieved is 10–29%;
L: Low performance if value achieved is 30–49%;
M: Moderate performance if value achieved is 50–69%;
H: High performance if value achieved is 70–89%;
E: Excellent performance if value achieved is 90–100%.

Table 10.9 summarizes this information for six different classifiers in three different metrics (Precision, Recall, and F1-Score) for the three classes (Converted, Demented, and Non-Demented). Note that a single classifier performing at the Excellent level in one metric for a certain class may not perform well in the same metric for a different class. In order to visualize the above data for a side-by-side comparison of the classifiers we assign a quality index for the performance. We use a six-point scale, assigning values of 1 through 6 for the performance levels of Scant through Excellent. For instance, Figure 10.8 shows how the different classifiers perform with respect to the different classes when Precision is used as the metric for the three different populations.

Similarly, Figure 10.9 compares the performance of the different classifiers when the metric used is Recall.

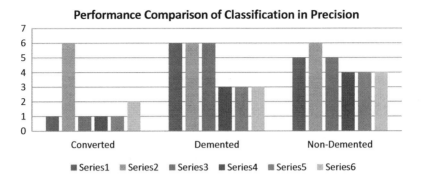

FIGURE 10.8
Performance Comparison of Classifiers in Precision.

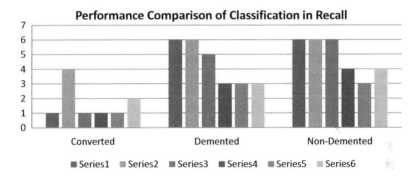

FIGURE 10.9
Performance Comparison of Classifiers in Recall.

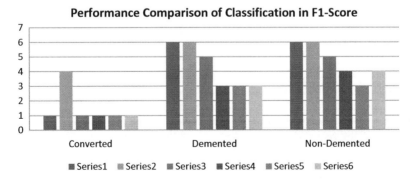

FIGURE 10.10
Performance Comparison of Classifier in F1-Score.

Finally, Figure 10.10 compares the performance of the different classifiers when using the F1-Score metric.

Note that the quality comparison of the classifiers when Precision is used as the metric suggests that RFC should be summarized in this information.

GBC and GPC are Excellent for Demented classification, whereas GBC alone is Excellent on Non-demented as well as on Converted. When the metrics of Recall and F1-Score are used, RFC and GBC exhibit excellent performance for demented and non-demented. On the other hand, on the classification of Converted, with the exception of GBC, these classifiers uniformly exhibit scant or poor performance. It is evident that GBC consistently outperforms its counterparts, in the classification of Converted, Demented, and Non-Demented, while NB, AdaBoost, and MLP are observed to perform far below average. Based on the results presented in the tables, we can also infer that GBC outperforms the rest of the classifiers. We believe that the excellent performance of GBC may be attributed to the process of bagging and boosting in GBC that ideally helps minimizing overfitting, bias, and variance.

10.5 Conclusion

Rather than finding a cure, there is more focus on risk reduction, early intervention, and timely diagnosis of dementia in older adults. As evident from the literature survey, a lot of work has been done for early detection of dementia using various machine learning algorithms and microsimulation techniques. However, there is still an utmost need for the identification of relevant attributes that could help detect dementia at an early stage.

References

[1] Tejada-Vera, B., 2013. Mortality from Alzheimer's disease in the United States; data for 2000 and 2010.

[2] Rueda, Andrea, Fabio A. Gonzalez, and Eduardo Romero. "Extracting salient brain patterns for imaging-based classification of neurodegenerative diseases." IEEE Transactions on Medical Imaging 33, no. 6 (2014): 1262–1274.

[3] Tohka, Jussi, Elaheh Moradi, Heikki Huttunen, and Alzheimer's Disease Neuroimaging Initiative. "Comparison of feature selection techniques in machine learning for anatomical brain MRI in dementia." Neuroinformatics 14, no. 3 (2016): 279–296.

[4] Chaplot, Sandeep, Lalit M. Patnaik, and N. R. Jagannathan. "Classification of magnetic resonance brain images using wavelets as input to support vector machine and neural network." Biomedical Signal Processing and Control 1, no. 1 (2006): 86–92.

[5] Herrera, Luis Javier, Ignacio Rojas, Alberto HéctorPomares, Olga Valenzuela Guillén, and Oresti Baños. "Classification of MRI images for Alzheimer's disease detection." In 2013 International Conference on Social Computing, pp. 846–851. IEEE, 2013.

[6] Li, Feng, Loc Tran, Kim-Han Thung, Dinggang Shen ShuiwangJi, and Jiang Li. "A robust deep model for improved classification of AD/MCI patients." IEEE Journal of Biomedical and Health Informatics 19, no. 5 (2015): 1610–1616.

[7] Ye, Jieping, Wu Teresa, Li Jing, and Kewei Chen. "Machine learning approaches for the neuroimaging study of Alzheimer's disease." Computer 44, no. 4 (2011): 99–101.

[8] Lama, Ramesh Kumar, Jeonghwan Gwak, Jeong-Seon Park, and Sang-Woong Lee. "Diagnosis of Alzheimer's disease based on structural MRI images using a regularized extreme learning machine and PCA features." Journal of Healthcare Engineering (2017).

[9] Torabi, Meysam, Reza Dehestani Ardekani, and Emad Fatemizadeh. "Discrimination between Alzheimer's disease and control group in MR-images based on texture analysis using artificial neural network." In 2006 International Conference on Biomedical and Pharmaceutical Engineering, pp. 79–83. IEEE, 2006.

[10] Torabi, Meysam, Hassan Moradzadeh, S. Reza Vaziri, Mohammad Javad Razavian, Reza Dehestani Ardekani, Moones Rahmandoust, Ali Taalimi, and Emad Fatemizadeh. "Development of Alzheimer's disease recognition using semiautomatic analysis of statistical parameters based on frequency characteristics of medical images." In 2007 IEEE International Conference on Signal Processing and Communications, pp. 868–871. IEEE, 2007.

[11] Huang, Chengzhong, Bin Yan, Hua Jiang, and Dahui Wang. "Combining voxel-based morphometry with artificial neural network theory in the application research of diagnosing Alzheimer's disease." In 2008 International Conference on BioMedical Engineering and Informatics, Vol. 1, pp. 250–254. IEEE, 2008.

[12] El-Dahshan, El-Sayed Ahmed, Tamer Hosny, and Abdel-Badeeh M. Salem. "Hybrid intelligent techniques for MRI brain images classification." Digital Signal Processing 20, no. 2 (2010): 433–441.

[13] Sampath, R., and A. Saradha. "Alzheimer's Disease Image Segmentation with Self-Organizing Map Network." JSW 10, no. 6 (2015): 670–680.

[14] Freund, Yoav, and Robert E. Schapire. "A decision-theoretic generalization of on-line learning and an application to boosting." Journal of Computer and System Sciences 55, no. 1 (1997): 119–139.

[15] Random Forest Simple Explanation. https://medium.com/@williamkoehrsen/random-forest-simple-explanation377895a60d2d. Accessed: 2019-01-04.

[16] Iddamalgoda, Lahiru, Partha S. Das, Achala Aponso, Vijayaraghava S. Sundararajan, Prashanth Suravajhala, and Jayaraman K. Valadi. "Data mining and pattern recognition models for identifying inherited diseases: challenges and implications." *Frontiers in genetics* 7 (136), pp 1-9, 2016.

[17] Grover, Prince. Gradient boosting from scratch. https://medium.com/mlreview/gradient-boosting-from-scratch-1e317ae4587d. Accessed: 2019-01-04.

[18] Simple guide to confusion matrix terminology. www.dataschool.io/simple-guide-to-confusion-matrix-terminology/. Accessed: 2019-01-04.

[19] Precision and Recall. https://en.wikipedia.org/wiki/Precision_and_recall. Accessed: 2019-01-04.

[20] F1 Score https://en.wikipedia.org/wiki/F1-Score. Accessed: 2019-01-04.

11

Fetal Standard Plane Detection in Freehand Ultrasound Using Multi Layered Extreme Learning Machine

S. Jayanthi Sree

Government College of Technology, Coimbatore, Tamil Nadu, India

C. Vasanthanayaki

Government College of Engineering, Salem, Tamil Nadu, India

11.1 Introduction

Ultrasound is the primary imaging modality for fetal health monitoring throughout the entire pregnancy since it is safe due to absence of harmful radiations, less expensive, and widely available. Abnormal fetal development is associated with an increased neonatal morbidity and mortality (Salomon et al. 2011). In most of the countries, a routine anomaly ultrasound scan is done around midpregnancy, that is, between 18 and 22 weeks. The anomaly scan is for imaging number of fetal organs and parts at the corresponding standard plan for measurements and also for detecting any fetal abnormalities (Salomon et al. 2011). A survey report in UK indicates that 83% of abnormalities incompatible with life, 50% of serious abnormalities where survival is possible, and 16% for those requiring immediate care after birth can be detected using ultrasound screening (Royal College of Obstetricians and Gynaecologists 2010). Identifying the standard planes for measurement is a tedious task since the quality of the ultrasound view of the fetus depends on a number of factors: artifacts in the ultrasound images (Zalud et al. 2009); body mass index (BMI) of the mother; position and motion of the fetus. Manual detection of standard planes of fetal parts and locating the specified fetal structure is time-consuming and expert dependent, requiring many years of experience for accurate results (Chan et al. 2009). Repetitive and lengthy examinations of the fetal structures cause stress resulting in inaccurate measurements. Automating the process of detecting the standard planes of fetal views for measurement is a relief to the busy schedule of the obstetricians since it saves time, has accuracy, robustness, reliability, and reproducibility (Zalud et al. 2009).

The automated standard plane detection can be used to provide real-time feedback about the particular frame to the freehand fetal ultrasound operator. The automated detection can also be used to retrieve standard planes of measurement from long videos for automated analysis of ultrasound examinations. The automated standard plane detection combined with the localization of structures helps in automated measurement of fetal parameters and anomaly detection. This also helps nonexperts and less trained operators in detection and diagnostic purposes with less error and higher reproducibility with reduced effort.

The chapter is organized as follows: Section 11.2 gives a survey of the existing works for fetal plane detection, Section 11.3 briefs about the ML-ELM architecture for fetal plane detection, and Section 11.4 gives the details of the experiments and discussion of the results.

11.2 Literature Review

A number of papers report various techniques for fetal structures detection in fetal 2D ultrasound examination videos. Yaqub et al. (2015) proposed guided random forests method for classifying 2D ultrasound images of fifth month pregnancy scans into seven standard scan planes. They modeled "other class" consisting of other background images which are not standard planes. Further works for fetal standard detection are based on Haar feature extraction and classifiers such as AdaBoost and random forests (Ni et al. 2013; Ni et al. 2014; Zhang et al. 2012). Chen et al. (2015a) detected the standard abdominal view using Convolutional Neural Network (CNN). The work was extended in Chen et al. (2015b) for detection of three standard planes. In these methods, the classifier had to be applied several times, thus increasing the processing time. Recently, deep neural networks based on CNN were developed (Baumgartner et al. 2017), where feature extraction and classification were performed as a unified network. Though they have shown higher accuracy, the computational cost during training and recognition process is expensive.

11.3 Multi Layered Extreme Learning Machine

11.3.1 Extreme Learning Machine Theory

Extreme Learning Machine (ELM) for Single Layer Feed forward Networks (SLFN) was proposed by Huang et al. (2006). The parameters of hidden layers of the ELM are randomly generated and need not be tuned. The input image is mapped to ELM random feature space of dimension N and the output is given by (11.1)

$$f_N(x) = \sum_{i=1}^{N} \beta_i k_i(x) = k(x)\beta \tag{11.1}$$

where $\beta = [\beta_1, \beta_2, ..., \beta_K]^T$ is the output weight matrix between hidden nodes and output nodes. The hidden node outputs for input x are $k(x) = [g_1(x), g_2(x),..., g_K(x)]$. The output of the ith hidden node is $g_i(x)$. ELM resolves the learning problem given in (11.2) for M training samples $\{(x_i, t_i)\}_{i=1}^{M}$

$$K\beta = T \tag{11.2}$$

where $T = [t_1, t_2,..., t_N]^T$ are the target labels and $K = [k^T(x_1), k^T(x_2),..., k^T(x_M)]^T$. The output weights are computed as given in (11.3)

$$\beta = K^{\Psi}T \tag{11.3}$$

where K^{Ψ} is the Moore–Penrose generalized inverse of matrix K. A regularization term (Huang et al. 2012) can be added to Equation (11.3) as given in (11.4)

$$\beta = \left(\frac{I}{C} + K^T K\right)^{-1} K^T T \tag{11.4}$$

11.3.2 Extreme Learning Machine Based AutoEncoder

Auto encoder is a feature extractor in a multilayer learning framework (Vincent et al. 2008). Extreme Learning Machine – Auto Encoder (ELM-AE) represents the input features meaningfully in three different representations: compressed, sparse, and equal dimension representations. The weights and biases of the hidden nodes which are chosen to be random are also selected such that they are orthogonal. The input data are represented meaningfully in a different dimension space by these orthogonal random weights as illustrated by Johnson–Lindenstrauss Lemma (Johnson et al. 1984) and given by (11.5).

$$k = g(lx + m); l^T = I; m^T m = 1 \tag{11.5}$$

where $l = [l_1, l_2, ..., l_K]$ are orthogonal random weights and $m = [m_1, m_2, ..., m_K]$ are orthogonal random biases between input nodes and hidden nodes.

The output weights are tuned to learn the transformation of the input data to the feature space and are calculated using (11.6).

$$\left. \begin{array}{c} \beta = \left(\frac{I}{C} + K^T K\right)^{-1} K^T T \\ \text{(Sparse \& Compressed Representation)} \\ \beta = K^T T; \beta^T \beta = 1 \\ \text{(For equal Dimension Representation)} \end{array} \right\} \tag{11.6}$$

Singular value decomposition is used to solve Equation (11.6) as given in (11.7).

$$k\beta = \sum_{i=1}^{N} v_i \frac{d_i^2}{d_i^2 + C} v_i^T X \tag{11.7}$$

where v is the eigen vectors of KK^T, d are the singular values of K related to data X.

11.3.3 Multi Layered Extreme Learning Machine

Multilayered Extreme Learning Machine (ML-ELM) (Kasun et al. 2013) is a simple stacked layer-by-layer architecture consisting of ELM-AE. The weights of each hidden layers of ML-ELM are initialized using ELM-AE performing unsupervised learning. The hidden layer activation functions of ML-ELM is linear, since for hidden layers the number of input and output nodes are equal. The activation function of the input layer to the first hidden layer is nonlinear since the input and output nodes are not equal. Regularized least squares are

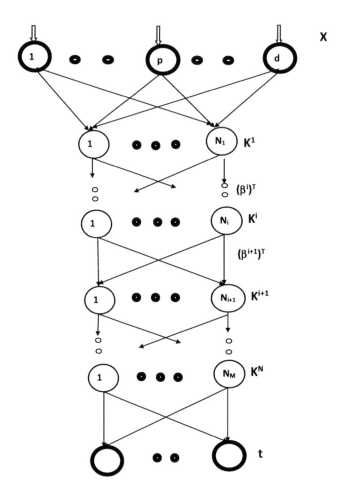

FIGURE 11.1
ML-ELM Network.

used to calculate the output node from the last hidden layer node. The network structure of ML-ELM is given in Figure 11.1 below. The first layer weights of ML-ELM are ELM-AE output weights β^1 with respect to input data x. The $(i + 1)$th layer weights of ML-ELM are the output weightsS β^{i+1} of ELM-AE, with respect to ith hidden layer output k^i of ML-ELM. Regularized least squares are used to calculate these weights.

11.4 Experimental Results

11.4.1 Dataset

The dataset for fetal standard plane detection consists of 15 ultrasound videos of gestational age, 18–22 weeks from two expert sonographers. Two different ultrasound systems

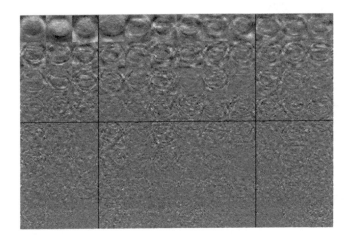

FIGURE 11.2
Input features represented by output weights β of ELM-AE.

CONFUSION MATRIX

150	31	19	70%
40	130	30	65%
11	13	176	88%
66.7%	70%	88%	74.1

OUTPUT CLASS — 1, 2, 3

1   2   3
TARGET CLASS

FIGURE 11.3
Confusion matrix of ML-ELM network.

of identical make (Philips) were used for acquisition. The frames from the images were split based on their corresponding standard view such as head, abdomen, and femur.

11.4.2 Experimental Setup and Preprocessing

The header of the ultrasound images containing machine setting and patient information was removed and any annotations in the images were also removed. All the frames images were rescaled to 200 × 200 images without any vendor logo and ultrasound control indicators. All the images were also normalized (subtraction of mean of the pixel intensities and division by the pixel intensities standard deviation). The dataset was split in the ratio of 80%:20% for training and testing, respectively. The splitting was done by standard plane level. The experiments were done in MATLAB R2017a on a desktop with a core i5-3470S CPU@2.90 GHz processor and 6 GB RAM.

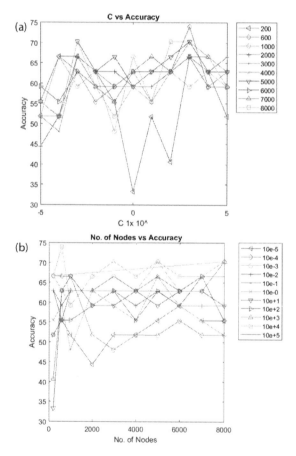

FIGURE 11.4

Accuracy of ML-ELM in terms of C and K (no. of hidden nodes): (a) C versus Accuracy and (b) no. of nodes versus Accuracy.

11.4.3 Discussion

The ML-ELM structure used in the experiment was: 40000-700-700-5000-3. The parameters for layer 40000-700 is 10^{-1}, for layer 700-5000 is 10^3, and for layer 5000-3 is 10^8 with sigmoidal activation for hidden nodes. The network gives a Testing accuracy of 75% in a testing time of 0.6 s. The Training time is about 6.02 s. The features which are represented by the output weights, β, are given in Figure 11.2. The confusion matrix of the ML-ELM network is given in Figure 11.3. The classes {1,2,3} represent standard planes of {Abdomen, Femur, and Head}.The user defined parameters, C, for regularized least mean square calculation and K, the number of hidden nodes are varied to find the dependency of Testing Accuracy over these parameters. The plot showing the variation of Testing accuracy with C subspace and K subspace is illustrated in Figure 11.4.

11.5 Conclusion

ELM-AE has good representation capability which could be useful in Muti-layer feed forward networks to provide better performance for fetal standard plane detection in terms of accuracy and processing speed. The method can be extended to focus on learning and detecting all standard planes of fetal structures and use of this technique for real-time fetal standard plane detection along with automatic fetal biometric measurements.

References

C.F. Baumgartner, K. Kamnitsas, J. Matthew, T.P. Fletcher, S. Smith, L.M. Koch, B. Kainz, and D. Rueckert. "SonoNet: Real-Time Detection and Localisation of Fetal Standard Scan Planes in Freehand Ultrasound." IEEE Trans Med Imaging 36 no. 11 (2017):2204–2215.

L.W. Chan, T.Y. Fung, T.Y. Leung, D.S. Sahota, and T.K. Lau. "Volumetric (3D) imaging reduces inter- and intraobserver variation of fetal biometry measurements." Ultrasound Obst Gyn 33 no. 4 (2009):447–452.

H. Chen, Q. Dou, D. Ni, J.-Z. Cheng, J. Qin, S. Li, and P.-A. Heng. "Automatic fetal ultrasound standard plane detection using knowledge transferred recurrent neural networks." Proceedings of MICCAI, Springer (2015a):507–514.

H. Chen, D. Ni, J. Qin, S. Li, X. Yang, T. Wang, and P.A. Heng. "Standard plane localization in fetal ultrasound via domain transferred deep neural networks." IEEE J Biomed Health Informatics 19 no. 5 (2015b):1627–1636.

G.-B. Huang, H. Zhou, X. Ding, and R. Zhang. "Extreme learning machine for regression and multiclass classification." IEEE Trans Syst Man Cybernet B Cybernet 42 no. 2 (2012):513–529.

G.-B. Huang, Q.-Y. Zhu, and C.-K. Siew. "Extreme learning machine: Theory and applications." Neurocomputing 70 (2006):489–501.

W. Johnson, and J. Lindenstrauss. "Extensions of Lipschitz mappings into a Hilbert space." Contemporary Mathematics Contem, 26 (1984):189–206.

L.L.C. Kasun, H. Zhou, G.-B. Huang, and C.M. Vong. "Representational learning with extreme learning machine for big data." IEEE Intell Syst 28 no. 6 (2013):31–34.

D. Ni, T. Li, X. Yang, J. Qin, S. Li, C.-T. Chin, S. Ouyang, T. Wang, and S. Chen. "Selective search and sequential detection for standard plane localization in ultrasound." International MICCAI

Workshop on Computational and Clinical Challenges in Abdominal Imaging, Springer (2013):203–211.

D. Ni, X. Yang, X. Chen, C.-T. Chin, S. Chen, P.-A. Heng, S. Li, J. Qin, and T. Wang. "Standard plane localization in ultrasound by radial component model and selective search." Ultrasound Med Biol 40 no. 11 (2014):2728–2742.

Royal College of Obstetricians and Gynaecologists. "Termination of pregnancy for fetal abnormality in England, Scotland and Wales." Report of a Working Party (2010).

L.J. Salomon, Z. Alfirevic, V. Berghella, C. Bilardo, K.-Y. Leung, G. Malinger, H. Munoz, et al. "Practice guidelines for performance of the routine mid-trimester fetal ultrasound scan." Ultrasound Obst Gyn 37 no. 1 (2011):116–126.

P. Vincent, H. Larochelle, Y. Bengio, and P.-A. Manzago. "Extracting and composing robust features with denoising autoencoders." Proceedings 25th Int. Conf. Mach. Learn., Helsinki, Finland (2008):1096–1103.

M. Yaqub, B. Kelly, A.T. Papageorghiou, and J.A. Noble. "Guided random forests for identification of key fetal anatomy and image categorization in ultrasound scans." Proceedings of MICCAI, Springer (2015):687–694.

I. Zalud, S. Good, G. Carneiro, B. Georgescu, K. Aoki, L. Green, F. Shahrestani, and R. Okumura. "Fetal biometry: A comparison between experienced sonographers, and automated measurements." J Mater Fetal Neonatal Med 22 no. 1 (2009):43–50.

L. Zhang, S. Chen, C.T. Chin, T. Wang, and S. Li. "Automated standard plane selection and biometric measurement of early gestational sac in routine ultrasound examination." Med Phys 39 no. 8 (2012):5015–5027.

12

Earlier Prediction of Cardiovascular Disease Using IoT and Deep Learning Approaches

R. Sivaranjani and N. Yuvaraj

Department of Computer Science and Engineering, KPR Institute of Engineering and Technology, Coimbatore, Tamil Nadu, India

12.1 Introduction

Health is all about the complete physical and mental strength and well-being and not only the absence of disease. Health care systems are designed to improve the health of people. In today's world, it is required to take an immediate action for health care. Preventive care in daily life has become more important against the increase in medical costs associated with aging and increase in lifestyle-related diseases [1]. Cardiac disease is the major reason for premature death worldwide [2]. It prevents the heart from fulfilling the circulatory demands since it impairs the ability to fill or eject blood from a ventricle.

An estimated 17.5 million people died from cardiovascular disease in 2012, representing 31% of global deaths. 17–45% patients die within the first year and the remaining die in the fifth year [3]. Based on the statistical report of American Heart Association (AHA), one out of three deaths is mainly due to cardiovascular disease. Cardiovascular disease is caused by a disorder of blood vessels in the heart. When the supply of blood to the heart is blocked, it often results in heart attack. This disease is associated with buildup of fats deposits in the arteries, thus increases risk. The factors that increase the risk of heart attack are high cholesterol level, diabetes, age, obesity, work stress, angina, previous heart attack, etc. It can be prevented by taking care of risk factors such as tobacco use, obesity an unhealthy diet, physical inactivity, and alcohol use. It is important to predict when a person is at high risk. Medical diagnosis needs to be carried out effectively and precisely. It would be better when it is automatically done. Internet of Things (IoT) and deep learning approaches can be used for the prediction of cardiovascular disorder.

Implementation of IoT can improve the quality of lives [4]. The future of the IoT is expected to have significant growth in home, healthcare, and business applications which increases the economic growth. Medical care and health care are the most attractive application areas in IoT [5] and it is considered as a backbone of IoT. IoT in health care gives support to fitness programs, chronic diseases, remote health monitoring, and elderly care. These services are expected to reduce costs, increase the quality of life, and enrich the user's experience [5]. IoT devices could be a part of the interconnected world. Sensors are connected to the devices called arduino or raspberry pi or pic microcontroller which are single-based microcontroller. Communication technologies enable IoT devices to communicate with smart devices. With the help of these components, data are collected and stored in an excel sheet. IoT is mainly used for collecting data from human or outside environment.

For the prediction of disease from the data collected, deep learning technique can be applied. Deep learning provides the methods to transform a large dataset into useful information for decision or prediction making. It is defined as an extraction of useful information from the data stored in a database [6]. The prediction is done based on the past training data. Deep learning can be successfully applied to big data for knowledge application, knowledge discovery, and knowledge-based prediction. It opens the door to enormous applications to find solutions for unsolvable problems with high speed. Application of mining in a medical field is used to find the patterns that are hidden in patient's medical datasets.

12.2 Literature Survey

Internet of Things is a new technology widely used nowadays. It makes things to communicate with each other through devices and the Internet. Implementation of IoT can improve the quality of lives [4]. The future of the IoT is expected to have significant growth in home, healthcare, and business applications, which increases the economic growth. Medical care and health care are the most attractive application areas in IoT [5] and it is considered as a backbone of IoT. The paper "Wearable wireless vital monitoring technology for smart health care" shows different health-related sensors that can be used and the relationship between use cases and applications of wireless system are also discussed. A patch-type wearable monitoring device is implemented with integrated sensors, a processor, and a Bluetooth transceiver. The kind of wireless system used usually depends on the type of sensor used. This device measures the living condition of users and their health status. The author explained how the sensors reduce human effort in collecting data. Communication technology is needed to store or transfer the data.

Deep learning is a well-developing research area in today's world and plays a major role in the development of the classification and predictive analysis systems. Enas M.F. El Houby analyzed the possible machine learning techniques to predict diseases. Their paper "A survey on applying machine learning techniques for management of diseases" used artificial neural network, k-nearest neighbor, decision tree, and associative classification for prediction purposes. This survey mainly focuses on choosing the best machine learning algorithm for disease prediction. The author had taken a private dataset and also data from different websites. When the data are extracted, machine learning algorithm is applied. He concludes by saying how important heart disease prediction is since it is the leading cause of death worldwide over the past 10 years. He compares each machine learning techniques based on accuracy, sensitivity, and specificity [6]. His survey answers the following questions: Which algorithm can be applied to manage the disease? What are the evaluation measures used to find the best algorithm?

In the paper, "Automated detection of cardiac arrhythmia using deep learning techniques," Swapna, Soman, and Vinayakumar discussed the different deep learning techniques for detecting arrhythmia. Dataset has been collected from MIT-BIH database in the PhysioNet [7]. The database has ECG recordings collected from different people. These recordings were digitized at a frequency of 360 data per second. Various trials were performed for choosing the optimal parameters. For analysis purpose, algorithm plays a major role since each algorithm produces different results. Convolutional neural network (CNN), recurrent neural network (RNN), long short-term memory (LSTM), and gated recurrent unit (GRU) and hybrid of CNN for automatic detection of abnormality were used. Keras and Tensorflow are used as a backend with graphics processing unit (GPU). This method does not need any feature extracting mechanisms and noise filtering.

In their paper, "Heart attack prediction system using cascaded neural network," Chitra and Seenivasagam used cascade neural network for prediction because they are self-organized and hidden layer can be increased during training phase. The system is designed for heart attack prediction using patient's medical record. Here, the dataset is collected from UCI Centre for Machine Learning. There are 76 attributes but their experiment refers to the use of 13 attributes. In order to remove duplicate records, cleaning and filtering is done followed by normalizing those values. Preprocessing is done initially using mining techniques and then attributes are classified using cascaded neural network. It takes less time to train the data. The accuracy of cascade neural network is compared with artificial neural network where the accuracy is 5% more than artificial neural network [8]. This accuracy can be varied based on the number of records taken.

Purushottama, Saxena, and Sharma analyzed the possibility and related matters of providing advanced services of human health management and a research direction of medical technology on IoT [9]. The main objective is to turn data into useful information and to make effective decision in medical field. The data were collected from the Cleveland Clinic Foundation. It contains 76 attributes but for experiments 14 attributes can be used. KEEL tool is preferred which is an open source tool to assess algorithms for solving problems. The paper "Efficient heart disease prediction system" uses classification decision rules in top-down approach. Classification decision rules include original rules, pruned rules, and classified rules. First, decision tree is generated to infer the rules. A hill climbing algorithm is also performed to find the best subset of rules. Accuracy for each rule has also been calculated [9]. Confusion matrix was built to find percentage of success for each partition.

Pandey (2017) uses both IoT and machine learning techniques for effective prediction of stress. The paper "Machine learning and IoT for prediction and detection of stress" uses health-related sensor and the different protocol for data collection. An efficient health monitoring system is designed and machine learning techniques like logistic regression and support vector machine are used for the detection of stress with the help of heartbeat. He also analyzed the possibility for checking whether the person is physically fit or unfit by using heartbeat of that person [10]. Based on training and testing accuracy, two different algorithms are compared. He provided an overview into the application of monitoring heart rate and this serves as a stepping stone to new research work in medical field.

Purushottam, Saxena, and Sharma designed a decision support system with Knowledge Extraction based on Evolutionary Learning (KEEL) tool, which is an open source (GPLv3) Java software tool to assess evolutionary algorithms for the desired problems. They collected data from the Cleveland Clinic Foundation. Missing values are filled by applying all possible-MV algorithm. This system generates rules such as original rules, pruned rules, classified rules, rules without duplicates, and sorted rules. The paper "Efficient heart disease prediction system using decision tree" also uses a special algorithm to fill the missing values in a dataset [11]. He found an accuracy of 87.3% in training and 86.3% in testing sets.

Kumar and Gandhi applied logistic regression with reduced dataset to develop the prediction model. An apache mahout framework has been used with elastic Map-Reduce framework for the development of prediction model in cloud computing. The paper "A novel three-tier Internet of Things architecture with machine learning algorithm for early detection of heart diseases" also evaluates performance by calculating sensitivity, specificity, and accuracy [12].

"Healthcare predictive analytics: An overview with a focus on Saudi Arabia" by Hana Alharthi briefly explained how predictive analysis for health care is more important and also discussed about the current efforts that are being implemented in developing countries. Logistic regression is used for predictive analysis since it provides confidence interval and odd ratio for each predictor in the model. She suggested the open source tool that includes rapid miner, Weka, KNIME and also the commercial data mining tools. She also discussed the challenges to implement predictive analysis in the healthcare sector. She suggested the dataset can be a patient's record digitized through EHR [13].

"Disease prediction by machine learning over Big data from Healthcare communities" by Chen, Hao, Hwang, Wang, and Wan designed a Convolutional Neural Network based Multimodal Disease Risk Prediction (CNN-UDRP) algorithm using data (both structured and unstructured data) from hospital. Here, the structured data include patient's information such as gender, age, etc. It also includes laboratory data. Unstructured data include doctor's interrogation records, narration of illness, and diagnosis [14]. The algorithms such as KNN, Naïve Bayesian, and Decision Tree are used to train structured data. For unstructured data, CNN-based unimodal disk prediction (CNN-UDRP) is used. While comparing with other algorithms, this proposed algorithm produces 94.8% accuracy.

Table 12.1 represents the comparative study between different algorithms used in related papers. Advantages and disadvantages for each algorithm are mentioned. From the above literature survey and comparison table (shown in Table 12.1), it is clear that

TABLE 12.1

Pros and cons of related papers.

Authors (Year)	Purpose	Methods	Pros and cons
Swapna G, Soman KP, Vinayakumar R (2018)	To Automatically detect the cardiac arrhythmia using deep learning techniques.	Convolutional neural network, recurrent neural network, long short-term memory and gated recurrent unit.	Automatic detection is done and for each technique accuracy has been calculated. But dataset is imbalanced.
Dr.T.Karthikeyan, V.A. Kanimozhi	To predict heart disease using deep learning approach.	Deep belief network, convolutional neural network	High accuracy is obtained. But only few samples are taken.
Enas M.F. El Houby (2017)	To make a survey on machine learning approaches to manage diseases.	Artificial Neural Network, K-Nearest Neighbour, Decision Tree, and Associative Classification	Accuracy is been calculated for each technique. But time-consuming process.
Purushottam, KanakSaxena, RichaSharma (2016)	To develop an efficient heart disease prediction system.	Classification decision rules	An efficient decision-making system with given parameters.

(*Continued*)

TABLE 12.1 (Cont.)

Authors (Year)	Purpose	Methods	Pros and cons
PurnenduShekhar Pandey (2017)	To develop a stress detection system using IoT and machine learning techniques.	Logistic regression and Support vector machine.	High accuracy is obtained for determining the characteristic of stress. But only fewer samples are taken.
Purushottam, KanakSaxena, RichaSharma (2015)	To design a decision support system with KEEL.	Decision tree	Can able to get an accuracy of 85%. But only fewer samples are considered.
PriyanMalarvizhi Kumar, Usha Devi Gandhi (2017)	To design a prediction model for earlier prediction of heart disease using machine learning approaches.	Logistic regression	It calculates sensitivity, specificity for each parameter. But the methodology is not clear.
Nam T. Nguyen (2017)	To develop an emotion prediction system using heart rate signals.	K Nearest neighbor, support vector machine	It can easily support any heart-rate sensor and Smartphone to enhance users' experience. But lower accuracy.
Monira Islam, Md.Ashikuzzaman, TahmidaTabassum and Md. Salah Uddin Yusuf (2017)	Prediction of heart disease from Photoplethysmography Signal.	A non-invasive technique (Independent component analysis)	It predicts heart disease froma facial video. Sometimes accuracy is very low.
Evanthia E. Tripoliti, Theofilos G. Papadopoulos, Georgia S. Karanasiou, Katerina K. Naka, Dimitrios I.Fotiadis (2016)	To present the state of the art of the machine learning methodologies applied for the assessment of heart failure.	K Nearest neighbor, support vector machine, random forests, Decision tree, neural network, clustering [20]	It focuses on all aspects of the management of heart failure. But the data set is highly imbalanced.
Min Chen, Yixue Hao, Kai Hwang, Lu Wang, and Lin Wan	To predict disease using machine learning algorithm over big data from healthcare communities.	Naïve Bayesian, Decision Tree, KNN	Faster than CNN-UDRP algorithm. But it takes more time to train the algorithm.
Julian Betancur et al.	To predict obstructive disease from fast myocardial perfusion SPECT using deep learning.	Model was implemented with the help of deep learning toolkit called Caffe and the training was done using graphical processor units(GTX 1080 Ti) [24].	It produces better results with higher accuracy but it has only limited amount of datasets.
Sanjay Purushotham, Chuizheng Meng, Zhengping Che, Yan Liu,	To benchmark the deep learning models on large healthcare datasets.	Deep learning models and ensemble of machine learning models [25].	Based on the time taken, the prediction is done. It uses large number of data and it makes sure that it can produce better results.
Bibo Shi, et al.	To predict the occult invasive disease in ductal carcinoma in situ using deep learning features.	Deep convolutional neural network model.	It shows an alternative way to analysis when the large datasets is not available. It needs prior domain knowledge for analysis.

UCI machine learning repository directory can be used. The UCI machine learning repository is the collection of domain theories and databases used by machine learning community. Deep learning algorithm based model is best suited for data classification and prediction purpose, because of its accuracy and it takes minimum time to train the model. The number of hidden layers used can be increased to get higher accuracy. It could solve the problems that arise in day-to-day life. To collect real-time data, IoT is preferred which uses sensor that can be used as a training dataset.

12.3 Methodology

This section includes how the data are collected using Internet of Things and how they are analyzed to make a prediction.

12.3.1 Data Collection Using IoT

IoT plays a major role while collecting real-time data. It mainly reduces the human effort. Type of sensors, type of microcontroller, and type of communication technology used are discussed in the following sections.

12.3.1.1 Sensors

Heartbeat sensor, temperature sensor, and respiratory sensor are used to collect real-time data. Heartbeat sensor is an electronic device that measures the heart rate (heartbeat per minute). For each beat, the LED flashes [13]. It normally works on the principle of light modulation by blood flow in a finger at each pulse. Temperature sensor is used in this model to measure the hotness of a body. LM35 temperature sensor is preferred because it has low self-heating. Respiratory sensor is used for measuring breathe rate (breaths per minute). The output from the sensor is in a digital format.

12.3.1.2 Microcontroller Board

A microcontroller board is a printed circuit board (PCB) combined with a processor, chipset, serial port, memory, SD card slot, Ethernet, etc. The open source tool that is best suited to work with the IoT is Arduino since it is an open source computer with flexible and easy-to-use software and hardware components [14]. It is designed by using various microprocessor and controllers. It works in a serial communication interface. They are preprogrammed with a bootloader and the programs are uploaded to on-chip flash memory. The main features of Arduino are 5 V (operating voltage), 8 bit, 16 MHz, very simple to program, and has analog I/O pins.

12.3.1.3 Communication Technology

The communication protocols provide an improved insight into the functionality and actual meaning of IoT [15]. Here, Zigbee is used since it is a wireless communication protocol, designed for short-term communication with low-energy consumption [16].

Zigbee protocol is created by Zigbee Alliance based on wireless IEEE802.15.4 [17]. The advantages of Zigbee are low cost, low data rate, low energy consumption, low complexity, reliability, and security. Zigbee network can support star, tree, and mesh topologies [16].

12.3.2 Working of IoT

Arduino board and PC are connected through data cable. Arduino software is installed and a program called Arduino c can be run for data collection. The sensors including heartbeat, respiratory and body temperature sensor periodically take a reading and send it. The reading can be seen in the Arduino software. Later it can be stored in excel sheet which can be used as a testing data. The collection of data using IoT is shown in Figure 12.1.

12.3.3 Training the Data Using Deep Learning Techniques

The dataset used and the different deep learning algorithm implemented are discussed in the following sections.

12.3.3.1 Dataset

A predictive analysis system can be developed from a training dataset. This is the actual data where the algorithm can be applied to train machine so that it can work automatically. From the test data, it can be found whether the machine learning algorithm reaches its accuracy. Training dataset can be collected from the UCI machine learning repository database [18]. It is the collection of datasets or databases that are used by deep learning community for analysis purpose. The dataset consists of parameters which tell about the patient details. The parameter includes name, age, sex, heart rate, body temperature, blood pressure, respiratory rate, and blood sugar level. Among these

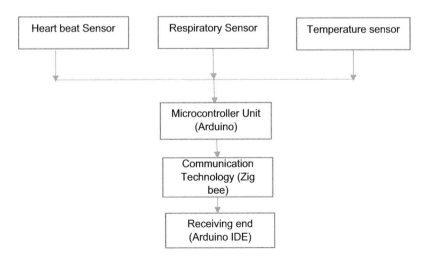

FIGURE 12.1
Data collection module in IoT (sensors and communication technology that are connected to obtain data).

attributes, heart rate and respiration play a major role in prediction. Among the dataset 70% of data is training data and 30% is test data. Following are the 17 attributes that are mentioned in Table 12.2.

12.3.3.2 Deep Learning Approach

Algorithm plays a major role in efficient decision-making process. From the comparative study, deep learning is preferred because it is robust in the noisy environment. The number of nodes can be increased to get higher accuracy. There are many algorithms that are available.

Deep learning algorithm can be applied for classification or prediction purpose using a technique called backpropagation. Deep neural network consists of an input layer, one or more hidden layers, and an output layer [19]. It is also an artificial neural network which has more than two hidden layers. Each layer is made up of units. Neural network can be defined as a set of connected input and output unit in which each connection has a weight. The inputs are continuously fed into the units making up the input layer. These inputs pass through the input layer and are then weighted and fed simultaneously to a second layer called a hidden layer. The outputs of the hidden layer units can be given as input to another hidden layer, and so on. The weighted outputs of the last hidden layer are input to units making up the output layer, which produces the network's prediction for given tuples. The input layer has units called input units. The units in the hidden layers and the output layer are called neurodes or output units [19].

Backpropagation is a type of neural network learning algorithm. Figure 12.2 represents the working model for backpropagation technique. To predict class label, the network learns by adjusting the weights. Artificial neural network is capable of learning

TABLE 12.2

Attributes and its range.

Attributes	Range
Age	>=20
Gender	F/M
Blood Pressure (mm Hg)	90/60 to 150/90
Cholesterol(LDL) (mg)	100 to 160
Heredity	YES/NO
Blood sugar (mg/dl)	80–125
PQ value	21.6–76
ST value	9.89–27.63
QT value	22.022–153
QRS value	8.532–68
R value	104–320
Heart beat rate	60–95
Respiratory rate	12 to 25 per minute
BMI	MEN < 26, WOMEN < 24
Smoking habit	YES/NO
Alcohol Intake	YES/NO
Mental stress	YES/NO

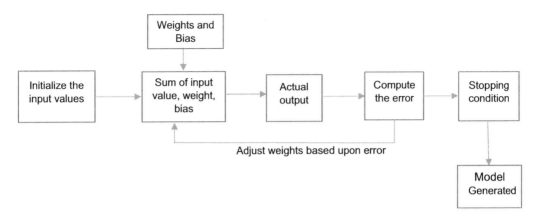

FIGURE 12.2
Working of backpropagation.

from examples and this system adapts itself during a training period, based on the existing examples of similar problems without the desired solution to each problem. After sufficient learning, the neural computer can be able to relate the problem data to the solutions, inputs to outputs, and it is then able to offer a viable solution to a problem.

12.3.3.3 Tool Used for Prediction

Pycharm is an integrated development environment (IDE) and one of the most powerful tools that can be used for data analytics. It is used in computer programming specially called python. Pycharm is preferred because it is a visual bugger which provides more options to debug python and JavaScript code and also it helps to do refactoring.

12.4 Architecture and Its Working

Medical decision support system has two modules. One is data collection and another one is analytics. Data collection module involves collecting data from database (that can be used as training data) and sensing devices (data from sensing device used as testing data). Analytical module involves analysis of data to obtain prediction. Training data and testing data are given as input in pycharm. These data are analyzed with the help of deep learning. Analytical module involves four steps which includes:

- Data gathering
- Data preprocessing
- Applying deep learning algorithm
- Knowledge discovery

Data gathering involves collecting data from machine learning repository which is a training data. It gathers all raw and reliable data. It is a time-consuming process but it produces credible results. After gathering data, preprocessing is done. Preprocessing involves converting raw data into an understandable format. The dataset contains invalid values and missing values. It is very difficult for a deep learning algorithm to process on these values. So, for preprocessing four steps have to be followed. They are:

- Cleaning
- Integration
- Transformation
- Reduction

Cleaning involves filling missing data and removing noisy data. Integration involves joining data from other sources. Transformation process normalizes the data, and also aggregation and generalization is done. The attributes which are least required are removed in reduction process.

After preprocessing, deep learning algorithm is applied. The main advantage of using deep learning is any number of hidden layers are included in order to get higher accuracy. Here five hidden layers are used. The working of deep learning network is summarized as follows:

- Input data are given to input layer, after processing it produces a certain output.
- The predicted output is subtracted from the actual output, later error value is calculated.
- Then backpropagation technique is used to adjust the weights.
- For weight adjustment, it starts from the weight between last hidden layer and output layer nodes and propagates backward through network.
- After finishing the backpropagation, it again starts the forwarding process.

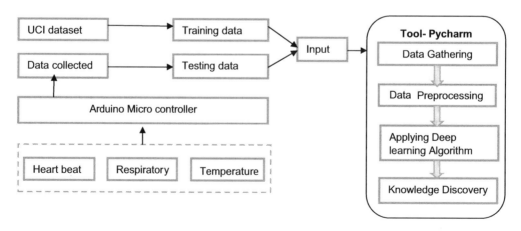

FIGURE 12.3
Medical decision support system (architecture).

- This process continues till the error between actual output and predicted output is minimized.

These algorithms are implemented in python. Analysis module randomly samples data to make sure that both training and testing sets are similar. Finally, some knowledge is gained from knowledge discovery phase. It predicts the disease when the person is in high risk. Cardiovascular disease is predicted from the dataset using medical decision support system represented in Figure 12.3.

12.5 Performance Measures

Confusion matrix (also known as an error matrix) is a table that allows visualizing the performance of an algorithm. Each row represents the instances in predicted class and each column represents the instances in actual class. Figure 12.4 represents the confusion matrix.

The performance of the algorithms can be evaluated by the following criteria:

- Speed: The computation cost involved to generate the model and to use the model.
- Scalability: The ability to construct an efficient model with the given dataset.
- Interpretability: The level of understanding of the model.
- Accuracy: The ability to predict the class label correctly (formula for accuracy is given in (12.1)).

$$Accuray = (TP + TN)/(TP + TN + FP + FN) \tag{12.1}$$

where TP implies true positive, TN implies true negative, FP implies false positive, and FN implies false negative.

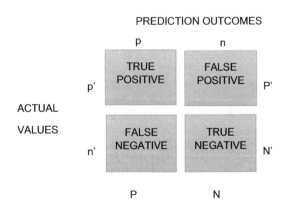

FIGURE 12.4
Confusion matrix (based on true positive, true negative, false positive, and false negative).

Other performance measures include sensitivity and specificity, as given in (12.2) and (12.3).

- Sensitivity: It is the measure of test's ability to detect condition when it is present.
- Specificity: It is the measure of test's ability to exclude the condition correctly when it is absent.

$$\text{Sensitivity} = TP/TP + FN \tag{12.2}$$

$$\text{Specificity} = TN/TN + FP \tag{12.3}$$

12.6 Conclusion

A model is presented for predicting cardiovascular disease using deep learning algorithm. Earlier prediction of a disease may lead to a cure or longer survival. Heart rate is one of the important parameters to check whether the person is physically fit or not. A real-time pulse rate/heart rate monitoring with heartbeat sensor provides an accuracy and stability in heart rate detection. After obtaining the parameters, the device (Arduino) will send data via Zigbee [21][22]. Prediction is done using datasets that include heart rate, temperature, and respiratory rate with the help of algorithm. It is a solution to overcome barriers to improve health care in low- and middle-income countries. The next step in this work would be, along with the heart rate, galvanic skin response used to detect the mood of a person. Also for more accurate results, more number of hidden layers can be used in the future.

References

[1] Arshdeep Bahga, Vijay Madisetti, "Internet of Things: A Hands-on Approach," Georgia Institute of Technology, Georgia, 2014.
[2] Nam T. Nguyen, Nhan V. Nguyen, T. My Huynh, Binh T. Nguyen Tran "A Potential Approach for Emotion Prediction Using Heart Rate Signals," International Conference on Knowledge and Systems Engineering, Vietnam, 2017 IEEE.
[3] Monira Islam, Md Ashikuzzaman, Tahmida Tabassum, Md. Salah Uddin Yusuf, "A Non-Invasive Technique of Early Heart Diseases Prediction from Photoplethysmography Signal," IEEE, Bangladesh, 2017.
[4] Ala Al-Fuqaha, Mohammed Aledhari, "Internet of Things: A Survey on Enabling Technologies Protocols, and Applications," IEEE, USA, 2015.
[5] Ferdoush, Xinrong Li, "Wireless Sensor Network System Design Using Raspberry Pi and Arduino for Environmental Monitoring Applications," Department of Electrical Engineering, University of North Texas, Denton, TX,2014.
[6] Enas M.F. El Houby, A Survey on Applying Machine Learning Techniques for Management of Diseases, National Research Centre, Systems and Information Department, Engineering Division, Cairo, Egypt, Elsevier, 2017.
[7] G. Swapna, K.P. Soman, R. Vinayakumar, "Automated Detection of Cardiac Arrhythmia Using Deep Learning Techniques", Elsevier, India, June 2018.

[8] R. Chitra, V. Seenivasagam, "Heart Attack Prediction System Using Cascaded Neural Network," Proceedings of the International Conference on Applied Mathematics and Theoretical Computer Science, India, 2013.

[9] Kanak Saxena Purushottama, Richa Sharma, "Efficient Heart Disease Prediction System," Elsevier, Madhya Pradesh, India, 2016.

[10] Purnendu Shekhar Pandey, "Machine Learning and IoT for Prediction and Detection of Stress," IEEE, Haryana, India, 2017.

[11] Purushottam Prof. (Dr.) Kanak Saxena, Richa Sharma, "Efficient Heart Disease Prediction System Using Decision Tree," IEEE, Madhya Pradesh, India, 2015.

[12] Priyan Malarvizhi Kumar, Usha Devi Gandhi, A Novel Three-Tier Internet of Things Architecture with Machine Learning Algorithm for Early Detection of Heart Diseases, Elsevier, India, 2017.

[13] Hana Alharthi, "Healthcare Predictive Analytics: An Overview with a Focus on Saudi Arabia," Journal of Infection and Public Health, Elsevier, Saudi Arabia, November 2018.

[14] Min Chen, Yixue Hao, Kai Hwang, Lu Wang, Lin Wan, "Disease Prediction by Machine Learning Over Big Data from Healthcare Communities," IEEE, China, 2017.

[15] Mobyen Uddin Ahmed, Shahina Begum, "Internet of Things (IoT) Technologies for Health Care," France, October 2017.

[16] Doaa Alrababah, Esraa Al-Shammari, Areej Alsuht, "A Survey Authentication Protocols or Wireless Sensor Network in the Internet of Things; Keys and Attacks," Princess Sumaya University for Technology, Amman, Jordan, 2017 IEEE.

[17] R. Sivaranjani, N. Yuvaraj, "Healthcare Access: A Survey on IoT and Machine Learning Algorithms for Effective Heart Rate Detection," IJSRD, Coimbatore, India, 2018.

[18] T. Karthikeyan, V.A. Kanimozhi, "Deep Learning Approach for Prediction of Heart Disease Using Data Mining Classification Algorithm Deep Belief Network," International Journal of Advanced Research in Science, Engineering and Technology, January 2017.

[19] http://myweb.sabanciuniv.edu/rdehkharghani/files/ 192016/02/The-Morgan-Kaufmann-Series-in-Data-Management-Systems-Jiawei-Han-Micheline-Kamber-Jian-Pei-Data-Mining.-Concepts-and-Techniques-3rd-Edition-Morgan-Kaufmann-2011.pdf

[20] Evanthia E. Tripoliti, Theofilos G. Papadopoulos, Georgia S. Karanasiou, Katerina K. Naka, Dimitrios I. Fotiadis, "Heart Failure: Diagnosis, Severity Estimation and Prediction of Adverse Events through Machine Learning Techniques," Elsevier, Greece, 2016.

[21] Gaurav Choudhary, A.K. Jain, "Internet of Things: A Survey on Architecture, Technologies, Protocols and Challenges," National Institute of Technology, Jalandhar, India, 2016 IEEE.

[22] Jie Lin, Wei Yuy, Nan Zhangz, Xinyu Yang, Hanlin Zhangx, Wei Zhao, "A Survey on Internet of Things: Architecture, Enabling Technologies, Security and Privacy, and Applications," 2017 IEEE.

[23] Julian Betancur, et al., "Deep Learning for Prediction of Obstructive Disease from Fast Myocardial Perfusion SPECT," The American College of Cardiology Foundation, Elsevier, California, 2018.

[24] Sanjay Purushotham, Chuizheng Meng, Zhengping Che, Yan Liu, "Benchmarking Deep Learning Models on Large Healthcare Datasets," Journal of Biomedical Informatics, University of Southern california, USA, 2018.

13

Analysis of Heart Disease Prediction Using Various Machine Learning Techniques

M. Marimuthu, S. Deivarani, and R. Gayathri
Coimbatore Institute of Technology, Coimbatore, Tamil Nadu, India

13.1 Introduction

In today's world people have a busy schedule which leads to stress and anxiety. The percentage of people with obesity and people addicted to smoking is increasing drastically. This leads to diseases like heart disease, cancer, etc. The challenge behind these diseases is its prediction. Each person has different pulse rate and blood pressure. The pulse rate must be between 60 and 100 beats per minute and the blood pressure must be in the range of 120/80 to 140/90. Heart disease is one of the major causes of death in the world (Benjamin et al., 2018). The number of people affected by heart diseases increases irrespective of age in both men and women. But other factors like gender, diabetes, and body mass index also contribute to this disease. In this chapter, we have tried prediction and analysis of heart disease by considering parameters like age, gender, blood pressure, heart rate, diabetes, and so on. Since numerous factors are involved in heart disease, the prediction of this disease is challenging. Some of the major symptoms of heart attack are:

  i. Chest tightness
 ii. Shortness of breath
iii. Nausea, indigestion, heartburn, or stomach pain
 iv. Sweating and fatigue
  v. Pain in the upper back that spreads to the arm.

Heart means "cardio." Hence, all heart diseases fall in the category of cardiovascular diseases. The different kinds of heart diseases are:

  i. Coronary heart diseases
 ii. Angina pectoris
iii. Congestive heart failure
 iv. Cardiomyopathy
  v. Congenital heart diseases (Kaur et al., 2016)

Coronary heart disease or coronary artery disease is the narrowing of the coronary arteries. The coronary arteries supply oxygen and blood to the heart. It is the most common type of heart disease leading to death. High blood glucose in diabetes patients can damage blood vessels and nerves that control the heart and blood vessels. If a person has diabetes for a longer time, there are high chances for that person to have heart disease in future. With diabetes, there are other reasons that contribute to heart disease. Smoking increases the risk of developing heart disease, high blood pressure makes the heart work harder to pump blood and it can strain the heart and damage blood vessels, abnormal cholesterol levels also contribute to heart disease and obesity. Also, family history of heart disease can be another cause, which is out of scope of this chapter.

The other risk factors include age, gender, stress, and unhealthy diet. Chances of having heart disease increase with age. Men have a greater risk of heart disease. However, women also have the same risk after menopause. Leading a stressful life can also damage the arteries and increase the chance of coronary heart disease.

So, in this chapter based on the factors mentioned before we try to predict the risk of heart disease. A large amount of work has been done related to prediction of heart disease by using various techniques and algorithms. These techniques may be based on deep learning, machine-learning, data mining, and so on. The aim of all those studies is to achieve better accuracy and to make the system more efficient to accurately predict the chances of heart attack.

13.2 Literature Review

Gandhi et al. used naïve Bayes (NB), decision tree, and neural network algorithms and analyzed the medical dataset. There are a huge number of features involved. So, there is a need to reduce the number of features. This can be done by feature selection. On doing this, time is reduced. They made use of decision tree and neural networks (Gandhi and Singh, 2015).

Thomas et al. made use of k-nearest neighbor (KNN) algorithm, neural network, naïve Bayes, and decision tree for heart disease prediction. They also made use of data mining techniques (Thomas and Theresa Princy, 2016).

Sana Bharti et al. made use of particle swarm optimization, artificial neural network, and genetic algorithm for prediction. Associative classification is a new and efficient technique which integrates association rule mining and classification to a model for prediction and achieving good accuracy (Bharti and Singh, 2015).

Purushottam et al., proposed:

> An automated system in medical diagnosis would enhance medical care and it can also reduce costs. In this study, we have designed a system that can efficiently discover the rules to predict the risk level of patients based on the given parameter about their health. The rules can be prioritized based on the user's requirement. The performance of the system is evaluated in terms of classification accuracy and the results shows that the system has great potential in predicting the heart disease risk level more accurately.
>
> (Purushottam et al., 2015)

Palaniyappan et al. made use of naïve Bayes, decision tree, and artificial neural networks to build intelligent heart disease prediction systems (IHDPS). To enhance visualization and ease of interpretation, it displays the results both in tabular and graphical forms. By providing effective treatments, it also helps to reduce treatment costs. Discovery of hidden patterns and relationships often has gone unexploited. Advanced data mining techniques helped remedy this situation (Palaniyappan and Awang, 2008).

Sharma et al. made use of decision tree, support vector machine (SVM), deep learning, k-nearest neighbor algorithms. Since the datasets contain noise, they tried to reduce the noise by cleaning and preprocessing the dataset and also tried to reduce the dimensionality of the dataset. They found that good accuracy can be achieved with neural networks (Himanshu Sharma and Rizvi, 2017).

Hazra et al., discussed in detail the cardiovascular disease and different symptoms of heart attack. The different types of classification and clustering algorithms and tools were used (Hazra et al., 2017).

Krishnaiah et al. presented an analysis using data mining. The analysis showed that using different techniques and taking different number of attributes gives different accuracies for predicting heart diseases (Krishnaiah et al., 2016).

Kaur and Kaur have showed that the heart disease data contain unnecessary, duplicate information. This has to be preprocessed. Also, they say that feature selection has to be done on the dataset for achieving better results (Kaur and Kaur, 2016).

Vijayashree et al. used data mining. A huge amount of data are produced on a daily basis. As such, it cannot be interpreted manually. Data mining can be effectively used to predict diseases from these datasets. In this paper, different data mining techniques are analyzed on heart disease database. In conclusion, this paper analyzes and compares how different classification algorithms work on a heart disease database (Vijayashree and NarayanaIyengar, 2016).

Benjamin et al. states that there are seven key factors for heart disease such as smoking, physical inactivity, nutrition, obesity, cholesterol, diabetes, and high blood pressure. They also discussed the statistics of heart disease including stroke and cardiovascular disease (Benjamin et al., 2018).

Kishore et al. showed that recurrent neural network gives good accuracy when compared to other algorithms like CNN, naïve Bayes, and SVM. Hence, neural networks perform well in heart disease prediction. They also achieved a system that could predict silent heart attacks and inform the user as earliest as possible (Kishore et al., 2018).

Nikhil Kumar et al. used various algorithms such as decision tree, random forest, naïve Bayes, KNN, support vector machine, and logistic model tree. Naïve Bayes algorithm gave good results when compared to other algorithms. They made use of UCI repository of heart disease dataset. Also, J48 algorithm took less time to build and gave good results (Nikhil Kumar et al., 2018).

Kaur and Arora compared various algorithms such as artificial neural network, k-nearest neighbor, naïve Bayes, and support vector machine on heart disease prediction (Kaur and Arora, 2018).

Weng et al. used four machine learning algorithms such as logistic regression, random forest, gradient boosting machines, and neural networks. They showed that machine learning algorithms perform well at predicting heart disease cases correctly. They say that this is the first experimentation using machine learning techniques to routine patient data in electronic records. The source of the dataset is the Clinical Practice Research Datalink (CPRD). These are the electronic medical records containing all the medical related data such as statistics of human population, medical history, and consultant specialists. It also contains details of medicine intake, outcomes, and details of hospital admissions (Weng et al., 2017).

Sahaya Arthy and Murugeshwari analyzed the existing works on heart disease prediction using data mining. Data mining techniques are commonly used in heart disease prediction. They also discuss the databases used such as the heart disease dataset from UCI repository; tools used such as Weka, Rapid Miner, Data melt, Apache Mahout, Rattle, KEEL, R data mining, and so on. They conclude that use of single algorithm results in better accuracy in prediction. But use of hybridization of two or more algorithms can enhance and improve the heart disease prediction with good accuracy (Sahaya Arthy and Murugeshwari, 2018).

Sudha et al. discusses the data mining technology. They also propose an architecture diagram which includes the following steps- dataset collection, normalization and preprocessing, dimensionality reduction using principal component analysis, feature subset selection, classification algorithm, and result analysis. They made use of three classifiers decision tree, naïve Bayes, and neural networks. They conclude that neural networks perform well than other classifiers (Sudha et al., 2012).

13.3 Heart Disease Prediction

In this chapter, comparison of various machine learning methods is done for predicting coronary heart disease of patients from their medical data in the next 10 years. Figure 13.1 will show the flowchart for proposed methodology.

The heart disease dataset is taken as input. It is then preprocessed by replacing nonavailable values with column means.

Four different methods were used in this chapter as depicted in Figure 13.2. The output is the accuracy metrics of the machine learning models. The model can then be used in prediction.

13.3.1 k-Nearest Neighbor

KNN is a nonparametric machine learning algorithm. The KNN algorithm is a supervised learning method. This means that all the data are labeled and the algorithm learns to predict the output from the input data. It performs well even if the training data are large and contain noisy values.

The data are divided into training and test sets. The training set is used for model building and training. A k-value is decided which is often the square root of the number of observations. Now the test data are predicted on the model built. There are different distance measures. For continuous variables, Euclidean distance, Manhattan distance, and Minkowski distance measures can be used.

However, the commonly used measure is Euclidean distance. The formula for Euclidean distance is as follows:

$$d = \sqrt{\sum\nolimits_{i=1}^{k} (x_i - y_i)^2} \qquad (13.1)$$

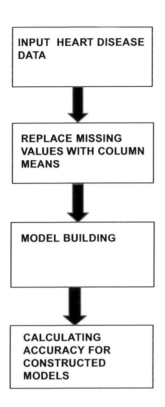

FIGURE 13.1
Proposed work.

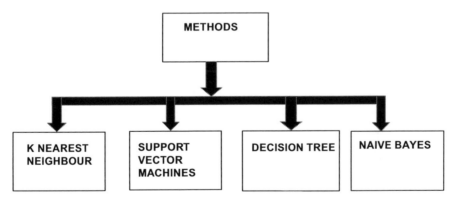

FIGURE 13.2
Methods used.

13.3.2 Support Vector Machine

The SVM algorithm is used to predict heart disease by plotting the training dataset where a hyperplane classifies the points into two presence and absence of heart disease.

SVM works by identifying the hyperplane which maximizes the margin between two classes.

Here, penalized SVM is used to handle class imbalance. Class imbalance is a problem in machine learning when total number of positive and negative class is not the same. If the class imbalance is not handled, then the classifier will not perform well. The class imbalance is shown in Figure 13.3.

SVM algorithms use a set of mathematical functions called kernel. In this proposed methodology, linear kernel is used.

$$k(x, \bar{x}) = \exp\left(-\|x - \bar{x}\|^2 / 2\sigma^2\right) \tag{13.2}$$

The performance of the SVM classifier can be increased by fine-tuning the hyperparameters. This can be done by using Grid Search CV. Different values of C can be given as input in this method. It builds different SVM models with given values and then finds the best value of C for which the model performs well.

13.3.3 Naive Bayes Algorithm

This is a classification algorithm which is used when the dimensionality of the input is very high. A naive Bayes classifier assumes that the presence of a particular feature in a class is unrelated to the presence of any other feature. It is based on Bayes theorem. The Bayes theorem is as follows:

$$P(A/B) = \frac{P(B/A) \cdot P(A)}{P(A)} \tag{13.3}$$

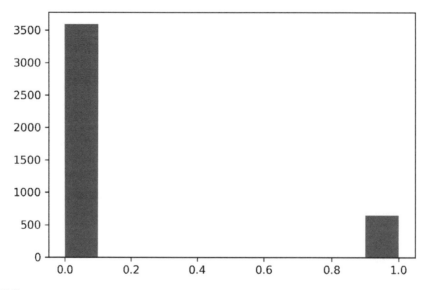

FIGURE 13.3
Plot showing class imbalance.

This calculates the probability of A and B where P(A) is the prior probability of A and P(B) is the prior probability of B.

It needs less training data. It can be used for binary classification problems and is very simple.

13.3.4 Decision Trees

Decision trees is a classic machine learning algorithm. In heart disease, there are several risk factors such as cigarette, blood pressure, hypertension, age, etc. The challenge of the decision tree lies in the selection of the root node. This factor used in root node must clearly classify the data. We make use of age as the root node.

The decision tree is easy to interpret. It is nonparametric and implicitly do feature selection.

13.4 Data Source

The dataset used is Framingham taken from Kaggle.

There were 16 attributes as follows: Male – gender 0 for female and 1 for male; Age – age of the patient; Education – values 1–5; education of the patient (values for those numbers not known); Current smoker – 1 if current smoker and 0 otherwise; Cigarette per day – if current smoker then number of cigarette per day; BP Meds – blood pressure; Prevalent BP – prevalent blood pressure; Prevalent Hyp – prevalent hypertension; Diabetes – 1 if diabetes 0 otherwise; Total cholesterol – cholesterol level; Sys BP – systolic blood pressure; Dia BP – diastolic blood pressure; BMI – body mass index; Heart rate – heart rate or pulse of the patient; Glucose – glucose level; Ten Year CHD – has chronic heart disease or not.

13.5 Results

The machine learning model is evaluated using the area under the curve-receiver operating characteristic (AUC_ROC) metric. This can be used to understand the model performance. The ROC curve of the algorithms is shown next.

The ROC curve is the receiver operating characteristic curve. The AUC is the area under the ROC curve. If the AUC score is high, the model performance is high and vice versa. Figures 13.4, 13.7 gives the ROC curve of the machine learning algorithms. The comparison of AUC score of various algorithms is given in Table 13.1. The accuracy of the algorithms is calculated. The accuracy results are tabulate in Table 13.2.

The accuracy of k-nearest neighbor algorithm is good when compared to other algorithms.

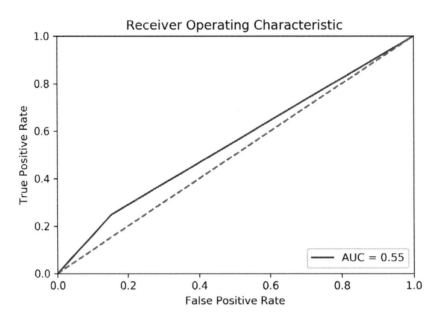

FIGURE 13.4
Receiver operating characteristic curve for decision tree.

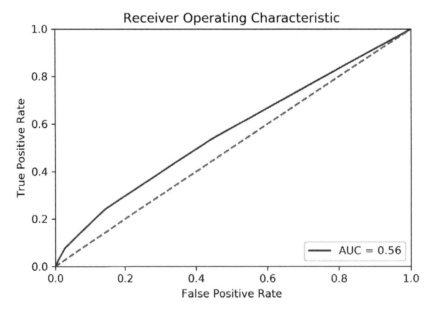

FIGURE 13.5
Receiver operating characteristic curve for k-nearest neighbor.

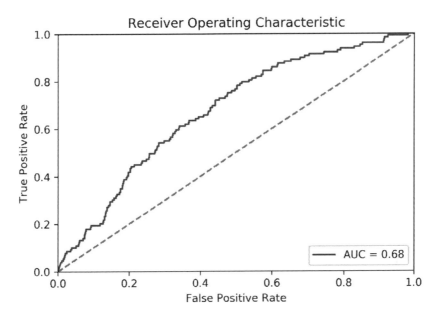

FIGURE 13.6
Receiver operating characteristic curve for naive Bayes.

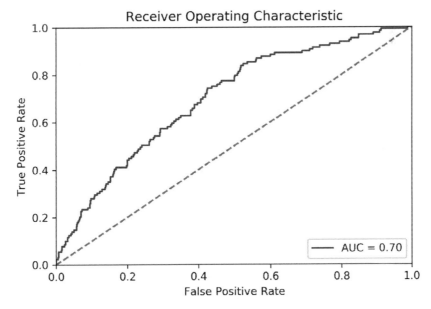

FIGURE 13.7
Receiver operating characteristic curve for support vector machine.

TABLE 13.1

Comparison of AUC scores of various algorithms.

Algorithm	AUC Score
Support vector machine	0.70
Naive Bayes	0.68
k-nearest neighbor	0.56
Decision tree	0.53

TABLE 13.2

Comparison of accuracy scores of various algorithms.

Method	Accuracy
k-nearest neighbor	83.60%
Naive Bayes	80.66%
Decision tree	75.58%
Support vector machine	65.56%

13.6 Conclusions and Future Work

This chapter discusses the various machine learning algorithms such as support vector machine, naïve Bayes, decision tree, and k-nearest neighbor which were applied to the dataset. It utilizes the data such as blood pressure, cholesterol, diabetes, and then tries to predict the possible coronary heart disease patient in the next 10 years.

Family history of heart disease can also be included for further increasing the accuracy of the model.

This work will be useful in identifying the possible patients who may suffer from heart disease in the next 10 years. This may help in taking preventive measures and hence try to avoid the possibility of heart disease for the patient. So when a patient is predicted as positive for heart disease, then the medical data for the patient can be closely analyzed by the doctors. An example would be, suppose the patient has diabetes which may be the cause for heart disease in future and then the patient can be treated for diabetes which in turn may prevent heart disease.

The heart disease prediction can be done using other machine learning algorithms. Logistic regression can also perform well in case of binary classification problems such as heart disease prediction. Random forests can perform well than decision trees. Also, the ensemble methods and artificial neural networks can be applied to the dataset. The results can be compared and improvised.

References

E.J. Benjamin et al., Heart Disease and Stroke Statistics 2018, At-a-Glance, 2018.

Sana Bharti and Shailendra Narayan Singh, Analytical Study of Heart Disease Prediction Comparing with Different Algorithms (May 2015), Amity University, Noida, India.

Monika Gandhi and Shailendra Narayanan Singh, "Predictions in Heart Disease Using Techniques of Data Mining." 1st International Conference on Futuristic Trend in Computational Analysis and Knowledge Management (ABLAZE-2015) (2015) 520–525.

Animesh Hazra, Subrata Kumar Mandal, Amit Gupta, Arkomita Mukherjee and Asmita Mukherjee, "Heart Disease Diagnosis and Prediction Using Machine Learning and Data Mining Techniques: A Review." Advances in Computational Sciences and Technology (2017) 2137–2159.

Amandeep Kaur and Jyoti Arora, "Heart Disease Prediction Using Data Mining Techniques: A Survey." International Journal of Advanced Research in Computer Science (2018) 569–572. Department of CSC Desh Bhagat University, Punjab, India.

Ramandeep Kaur and Prabhsharn Kaur, "A Review-Heart Disease Forecasting Pattern Using Various Data Mining Techniques." International Journal of Computer Science and Mobile Computing, Vol. 5, No. 6 (June 2016) 350–354.

Abhay Kishore, Ajay Kumar, Karan Singh et al., "Heart Attack Prediction Using Deep Learning." International Research Journal of Engineering and Technology (IRJET) (2018) 4420–4423.

V. Krishnaiah, G. Narsimha and N. Subhash Chandra, "Heart Disease Prediction System Using Data Mining Techniques and Intelligent Fuzzy Approach: A Review." International Journal of Computer Applications, Vol. 136, No. 2 (February 2016) 43–51.

M. A. Himanshu Sharma and A. Rizvi, "Prediction of Heart Disease Using Machine Learning Algorithms: A Survey." International Journal on Recent and Innovation Trends in Computing and Communication (August 2017) 99–104.

M. Nikhil Kumar, K.V.S. Koushik and K. Deepak, "Prediction Heart Diseases Using Data Mining and Machine Learning Algorithms and Tools." International Journal of Scientific Research in Computer Science, Engineering and Information Technology, Vol. 306 (2018) 888–898. Department of CSE, VR Siddhartha Engineering College, Vijayawada, Andhra Pradesh, India.

Sellappan Palaniyappan and Rafiah Awang, "Intelligent Heart Disease Prediction Using Data Mining Techniques." IJCSNS International Journal of Computer Science and Network Security, Vol. 8, No. 8 (August 2008) 343–350.

Purushottam, Kanak Saxena and Richa Sharma, "Efficient Heart Disease Prediction System Using Decision Tree." (2015) 72–77.

A. Sahaya Arthy and G. Murugeshwari, "A Survey on Heart Disease Prediction Using Data Mining Techniques." International Journal of Engineering Research in Computer Science and Engineering (IJERCSE), Vol. 5, No. 4 (April 2018) 280–287.

A. Sudha, P. Gayathri and N. Jaisankar, Effective Analysis and Prediction Model for Stroke Disease Using Classification Methods (April 2012).

J. Thomas and R. Theresa Princy, "Human Heart Disease Prediction System Using Data Mining Techniques." International Conference on Circuit, Power and Computing Technologies [ICCPCT] (2016).

J. Vijayashree and N. Chandra Sriman NarayanaIyengar, "Heart Disease Prediction System Using Data Mining and Hybrid Intelligent Techniques: A Review." International Journal of Bio-Science and Bio-Technology, Vol. 8, No. 4 (2016) 139–148.

Stephen F. Weng, Jenna Reps, Joe Kai, Jonathan M Garibaldi and Nadeem Qureshi, "Can Machine Learning Improve Cardio Vascular Risk Prediction Using Routine Clinical Data?" (2017) 1–14.

www.kaggle.com/amanajmera1/framingham-heart-study-dataset

14

Computer-Aided Detection of Breast Cancer on Mammograms

Extreme Learning Machine Neural Network Approach

Jayesh George M.
Vimal Jyothi Engineering College, Chemperi, Kerala, India

Perumal Sankar S.
Toc H Institute of Science and Technology, Kochi, Kerala, India

14.1 Introduction

Breast cancer is considered as a major reason of mortality among adult female. According to the National Cancer Institute Annual Report to the Nation 2017, incidence summary shows that between 1999 and 2013 the overall cancer incidence rate remained stable for women while the incidence rate continued to decrease among men. This type of cancer occurs almost entirely in women but in some rare cases men suffers from it too. There are about 20 different types of breast cancer. Most cancer occurs in the milk ducts and some in the glands.

The two types of tumors seen in women are benign and malignant. A noncancerous tumor is called benign tumor and is considered to be completely curable. A malignant tumor may invade the surrounding tissues and spread all over the body. The exact reason for breast cancer is still unknown but some of these are due to genetic abnormality and about 5–10% of cancers are inherited from parents. An X-ray imaging technique to examine human breast is called mammography. This specialized imaging technique aids in the early identification and diagnosing of breast abnormality as a screening tool. A diagnostic mammography is done for the patient who has previous abnormality and require some follow-up. A typical mammogram involves two or four views taken from different angles. A top view of breast is called cranial caudal view while a side view is called mediolateral oblique view.

Dense tissue and overlap of cancer cells with normal tissues leads to missed rate at the range of 10% in mammography. Interpretation of mammogram images is difficult because normal breast looks different for each women. It is necessary to have an advanced system between mammogram image reader and an input image to correctly identify the lesions. A computer-aided detection (CAD) system highlights the abnormal areas (mass, density, and microcalcification) on the images. [1] It will lead to a higher "recall" rate with less or no effect on positive predictive value for clinical biopsy. By improving the currently available CAD system it is possible to predict accurately the suspicious lesion so that rate of breast cancer recovery can be increased. [2]

A large number of researchers conducted study on CAD system for breast cancer and developed smart techniques to precisely identify and classify the cancer in last few years. [3–5] Number of studies has explained that computer-based detection of breast malignancy can ameliorate the identification accuracy rate from 4.6% to 19.8% compared to manual classification. A number of machine learning methods to classify samples as abnormal and normal have been presented for classification of abnormalities.

Karahaliou et al. [6] look into multiscale texture features of the skin around microcalcifications for breast malignancy diagnosis using probabilistic artificial neural network. Giger and Kupinski [7] explained a radial gradient index related method and a probabilistic algorithm for identification of cancer in digital X-ray images of breast. Sahiner et al. [8] studied with convolution neural network (CNN) classifier to discriminate the abnormality and the normal mammary tissue. Eltonsy et al. [9] developed a method based on the availability of concentric layers around the focal area with doubtful morphological features and less relative incidence in the breast area. Zheng et al. [10] proposed a multiscale feature using artificial neural network for identification of microcalcification distribution in digital breast images. All features derived are analyzed in spatial domain and use spectral entropy as a decision entity. Backpropagation with Kalman filtering (KF) is effectively utilized to obtain more accurate and efficient learning as required for checking of different input images and features.

Valarmathi et al. [11] developed an algorithm for malignancy identification based on feature extraction from different extraction methods such as discrete cosine transform along with Gabor filter, gray level spatial dependence matrix (GLSDM), and features are classified using artificial neural network. Angayarkanni and Nadira Banu Kamal [12] proposed a computer-based segmentation method of the breast images and classified them as benign, malignant, or normal with the help of the decision tree algorithm. A hybrid method of multiscale data mining was used to identify the texture features. The amount, distribution, and the stages of the malignancy were identified using the ellipsoid volume formula which was calculated over the segmented breast image region.

14.2 Methodology

The detailed methodology is given in Figure 14.1. We used MIAS database images for the experiments. A windowing function of size 32×32 is used to extract subimages from the breast images which are preprocessed. Sets of 20 GLSDM texture features, energy, and norm of Gabor filter and SURF filter coefficients are used to support the evaluation criteria of the suggested method. Section 14.3 explains the details of pre-processing methods. Section 14.4 gives details of breast image texture analysis. Section 14.5 describes extreme learning machine concepts and working. Section 14.6 shows performance of the system and results of proposed method. Section 14.7 discusses the conclusion drawn from the proposed method.

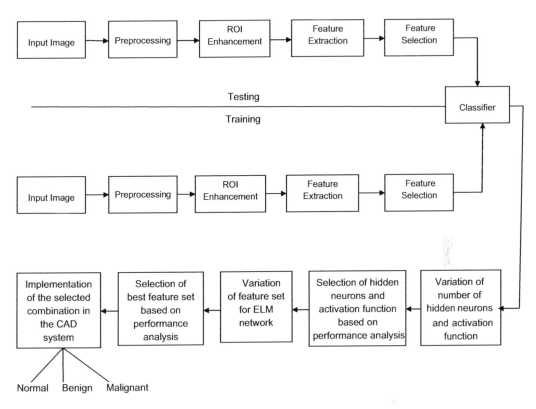

FIGURE 14.1
Proposed block diagram.

14.3 Preprocessing

The first step to any image processing technique is preprocessing as it improves the visual properties of the image. The basic aim of this method is to improve the visual appearance of the image with smoother unwanted artifacts which is important for next stage of evaluation. Preprocessing is essential as breast X-rays are very hard to analyze. [13]

14.3.1 Wiener Filter

Wiener filter is considered as the best method to reduce the general mean square error during smoothing of noise and inverse filtering. Wiener filter provides a linear prediction of actual input picture and least mean square error. The main aim of this filtering technique is to find a statistical estimate of unknown signal which will utilize an associated signal as a reference and filtering that signal to figure out the estimate of the final result. Wiener filter is mainly used to remove additive noise, which has constant spectral density, and reverse the obscuring simultaneously. Wiener filtering is superior to other methods in removing all visual artifacts. This filter gives the accurate technique for reducing the unarranged areas, with a precise aim to get the most suitable renewing of the original signal. [14] With reference to the psycho-visual criterion, most of the X-ray readers affirmed that Wiener filtering doesn't alter the information content of mammogram images, but it gives superior visibility. [15]

14.3.2 Breast Region Segmentation

The global threshold method is used to convert 2D gray images into black and white format, having a value 1/10 of the maximal intensity level. The black and white images with connected components are labeled. Connected component's area is measured and objects with number of pixels less than the maximum TA are eliminated. [16] The TA for MIAS dataset breast X-ray images is evaluated to be 10,000. The original input image is multiplied with breast profile to avail the image without label and any artifacts. [17]

After the segmentation, subimages (region of interests (ROIs)) of size 32 × 32 are wisely chosen to get the resultant image. The subimage size plays a vital role in classification of calcification with different features. A big subimage leads to proper identification of microcalcification in the homogenous area but a worst identification on breast border area, whereas smaller subimages provide reasonable identification correctly in most of the breast areas. The smaller and feasible subimage size that could better identify microcalcification tissues in a 1024 × 1024 X-ray image is 32 × 32, which is found out by trial and error method. In the regular breast X-ray images, areas of size 32 ×32 are considered as ROIs randomly. In areas affected by microcalcifications, region of interests are selected in a manner that the microcalcification distributions are at diverse areas of the cropped sub-X-ray image. [18] The black background subimages and images with zero count higher than 100 (10% of total count of pixel) are not used for identification. 120 region of interests are selected from 55 X-ray images (set of 25 microcalcifications and 30 normal is used for the experiment) to make learning and testing sets in which 86 region of interests (43 microcalcification and 43 normal) are chosen to make learning set and 34 region of interests (17 microcalcification and 17 normal) are chosen to test the performance.

14.4 Texture Analysis

Microcalcifications are primarily being studied with the help of texture analysis. [13,14] Texture analysis becomes an inevitable method to study microcalcification. This research has taken the advantages of texture analysis in order to detect particular region of interest, that is, microcalcifications. Variations in spatial distribution and related data could be identified through an analysis of textural features. The pixel intensity in a specific region contains the scheme of tone, for instance, gray level intensities in gray image. Hence, the pattern of variations in gray level values is explained through the texture characteristics of the cropped image. So, the necessity of textural features while ruling out the classification issues on nonhomogenous data is established already.

14.4.1 GLSDM Feature Extraction

The statistical way of analyzing the textures that deals with the spatial relationship of image pixel is done with GLSDM. The GLSDM distinguishes image texture by computing how often every pairs of pixels with specific value and the occurrence of specific spatial relationship in a breast image, generate a GLSDM, and subsequently bringing out statistical relation from this matrix. Gray level spatial dependence matrix is a competent statistical method for taking out second order data texture features from pictures. [19–21] The gray level spatial distribution in a breast image is conveyed through the GLSDM. Precisely, an entity in the gray level spatial dependence matrix, Px,y(i,j) stands for the probability of occurrence of the pairs of the gray levels (i,j) spaced by distance x at direction y. In this method, gray level spatial dependence matrix is calculated in two different directions y = 45 with pixel distance of two (x = 2). The different features used for the classifications are: difference variance, difference entropy, information measure of correlation, cluster prominence, dissimilarity, energy, entropy, sum entropy, inverse difference, inverse moment difference, normalized inverse difference, maximum probability, sum of squares-variance, average sum, variance sum, homogeneity, autocorrelation, contrast, correlation, and cluster shade.

14.4.2 Feature Vector Extraction Using Gabor Filter

Gabor filters are type of filters which effectively pass band of signals. Gabor filters have both frequency and orientation selective characteristics and both frequency and spatial domains have better joint resolution. [22–24] The texture information can be generated with the help of properly tuned Gabor filter to mammogram images. This texture information is usable to generate feature vector. The general form of Gabor filter in the spatial domain is given by [25]

$$h(x, y, \theta, f) = \exp\left\{ -\frac{1}{2}\left[\frac{x_\theta^2}{\sigma_\theta^2} + \frac{y_\theta^2}{\sigma_\theta^2}\right] \right\} \cos(2\pi f x_\theta) \tag{14.1}$$

where $x\theta = x\cos\theta + y\sin\theta$ and $y\theta = -x\cos\theta + y\sin\theta$

The parameters σ_x and σ_y of Gabor filter are modeled by a frequency f along x_θ direction.

The frequency is set by experimental study; the frequency f is selected as 7 pixel/cycle (1/7).

The value of θ can be written as

$$\theta_k = \frac{\pi(k-1)}{m}, \quad k = 1, 2, \ldots, m \tag{14.2}$$

Here, m refers to the count of orientations (m is selected to be 8 in this chapter). For an image of size $M \times N$ centered at (X,Y), with W, the Gabor magnitude is calculated [26] for $k = 1,2, \ldots, m$ as

$$g(X, Y, \theta_k, f, \sigma_x, \sigma_y) =$$
$$\sum_{x_0=M/2}^{(M/2)-1} \sum_{y_0}^{(N/2)-1} I(X + x_0, Y + y_0) h(x_0, y_0, \theta_k, f, \sigma_x, \sigma_y) \tag{14.3}$$

14.4.3 SURF Detector Based Feature Extraction

Speeded up robust features (SURF) is used for confined feature detection. Multiresolution pyramid method is used to convert the image into coordinate system. This is done to copy the original image with Laplacian Pyramid or Pyramid Gaussian shape to acquire a picture with equal size and limited bandwidth. A specific blurred view image is attained with Scale Space which assures that the interest points are scale and space invariant. This feature extraction method applies square-shaped filter as Gaussian smoothing approximation. When the integral image is applied, image filtration with a square is more rapid.

$$I(x,y) = \sum_{i=0}^{x} \sum_{j=0}^{y} S(i,j) \qquad (14.4)$$

Assessment of the sum of images in a rectangle could be done smoothly with the integral image that needs evaluations at the rectangle's four edges. The interested points can be detected at various scales, somewhat because the selection for correspondence sometimes needs images to compare which is usually evident at different scales. The Scale Space is generally perceived as an image pyramid in another feature extraction method. Firstly, Gaussian filter gets the images smoothed and then in order to attain the next higher level of the pyramid, the images undergo subsampling. Consequently, several floors or stairs with various sizes of masks are calculated.

$$\sigma_{approx} = \text{Current size of filter} * \left(\frac{\text{filter base scale}}{\text{filter base size}} \right) \qquad (14.5)$$

The space scale is subdivided into a number of octaves. Octave mentions a series of response maps dealing with a doubling of scale. In SURF, the lowest level of the scale space is achieved with the help of output of the 9×9 filters.

14.4.4 Support Vector Machine

Support vector machine (SVM) is used for classification and regression, and it is a machine learning algorithm based on the principle of structural risk minimization. SVM is mostly used for classification purpose. Based on the decision plane concept, this divides data into subset such that each element in the subset has similar characteristics. The SVM algorithm creates a line for the classification process and very effective for high-dimensional space. It can be used for both linear and non-linear problems.

For linearly separable binary set, SVM uses two sets of features. Suppose the two features are ×1 and ×2. We are using two classes—class triangle and class rectangle. The training vectors are classified into two classes by using a hyperplane that is designed by using the SVM. The classification model of SVM is shown in Figure 14.2.

There are different hyperplanes that will classify the vectors and the one with the maximum margin from both classes will be selected to classify these vectors. The margin is defined as the distance between the hyperplane and the closest element from this hyperplane. The hyperplane can be defined by using Equation (14.6)

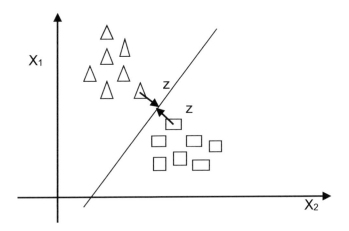

FIGURE 14.2
Classification using SVM.

$$g(\vec{x}) = w^{-T}\vec{x} + \omega_0 \tag{14.6}$$

where ω_0 is the vector of weight and if this equation has value greater than 1 for all the input vectors, then it represents class 1, that is, the class triangle. If the value of the equation is less than −1 for all the input vectors, then it belongs to class 2, that is, the class rectangle. This can be represented by Equations (14.7) and (14.8)

$$g(\vec{x}) \geq 1, \quad \forall \vec{x} \in class\ 1 \tag{14.7}$$

$$g(\vec{x}) \leq -1, \quad \forall \vec{x} \in class\ 2 \tag{14.8}$$

The distance to the closest element will be at least 1 and the distance between the point and the hyperplane can be computed. The aim is that minimizing this total distance will maximize the separability. Nonlinear optimization task is used to minimize this weight vector that will split these two classes, which can be solved by using Karush–Kuhn–Tucker (KKT) condition using Lagrange multipliers. The main equation states that the value of omega will be the solution of the sum in Equation (14.9).

$$\vec{\omega} = \sum_{i=0}^{N} \lambda_i y_i x_l \tag{14.9}$$

$$\sum_{i=0}^{N} \lambda_i y_i = 0 \tag{14.10}$$

There is also another rule that when we solve these equations trying to minimize the omega vector, we will maximize the margin between two classes and this will maximize the separability of two classes.

14.5 Extreme Learning Machine

Extreme learning machine (ELM) classifier is an effective, better, and time-conserving (for training) classifier with large number of uses in classification problem. The ELM classifier achieves good classification accuracies on proposed hyperspectral image classification on hyperspectral information sets when compared with traditional support vector machine. The dimensionality reduction is performed with semisupervised method. It is used to keep the Hughes phenomenon from frequently occurring in identification of large dimensional data. A progressive dimensionality trimmed down technique is used to mitigate the dimensionality of hyperspectral information. The aim of this method is to maximize the global functions which are calculated using the expression shown in Equation (14.11).

$$J = JD + \beta \bullet JR \tag{14.11}$$

Here, JR is the regularization parameter, JD is the discrimination term, and β is the parameter term. By applying PCA criteria as the normalizing term, Equation (14.6) can be rewritten as Equation (14.7), which is the global function of SSDRpca. Therefore, SSDRpca can be considered as specific idea of the semisupervised dimensionality mitigation method.

$$J_{SSDRpca}(w) = \frac{1}{2n_C} \sum_{(i,j)\in C} (w^T x_i - w^T x_j)^2 - \frac{\alpha}{2n_M} \sum_{(i,j)\in M} w^T x_i - w^T x_j)^2$$
$$+ \frac{\beta}{2N^2} \sum_{i,j} (w^T x_i - w^T x_j)^2 \tag{14.12}$$

By considering sparse representation as the normalization parameter, Equation (14.11) can be changed to Equation (14.13). The equation represents the global objective function of SSDRsp. The sparse reconstructive weight vector of x_i is represented by the term s_i.

$$J_{SSDRsp}(w) = \frac{1}{2n_C} \sum_{(i,j)\in C} (w^T x_i - w^T x_j)^2 - \frac{\alpha}{2n_M} \sum_{(i,j)\in M} (w^T x_i - w^T x_j)^2$$
$$+ \beta \bullet \left[-\frac{1}{N} \sum_i \left\| w^T x_i - w^T s_j \right\|^2 \right] \tag{14.13}$$

Extreme learning machine is designed for single hidden layer based feed forward artificial neural networks (SLFNs) and then extended to generalized feed forward network [27]. Instead of learning the hidden layer, ELM works on the principle of Moore–Penrose generalization. For a given N arbitrary distinct points (x_i, t_i), where input variables are $x_i = [x_{i1}, x_{i2}, ..., x_{in}]^T \in Rn$ and target values are $ti = [t_{i1}, t_{i2}, ..., t_{im}]^T \in R_m$, the output of SLFNs with $\overline{M}$ hidden nodes are mathematically represented as

$$f(y) = \sum_{i=1}^{\overline{M}} \beta_i \bullet g_i(y) \tag{14.14}$$

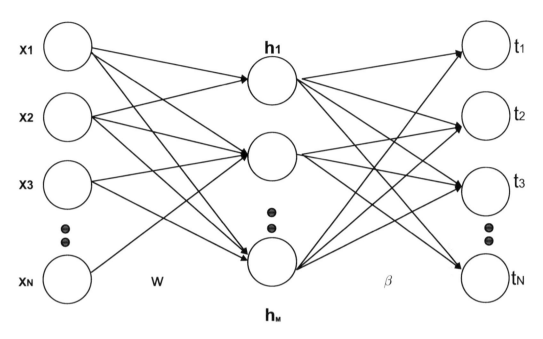

FIGURE 14.3
ELM network.

where $\beta_i \in R_m$ is the weight of the vector connecting the ith hidden node and the output layer nodes, and $g_i(y)$ is the activation function. The expression (14.14) could be rewritten after simplification as

$$H = T \qquad (14.15)$$

Here, $H_{M \times \overline{M}}$ is known as the hidden layer output matrix of the single hidden layer based feed forward artificial neural networks. The above linear system can be simplified by mathematical deductions. The smallest solution can be written as: $\hat{\beta} =$ H†T, where H† is the Moore–Penrose inverse of H. The ELM network structure is shown in Figure 14.3.

14.6 Performance Evaluation and Result

The implementation of extreme learning machine in the location of microcalcification is noticed to be the good choice compared to other well-known mainstream classifiers, namely, Naïve Bayes classifier, K Nearest Neighbor(KNN) and Support vector Machine (SVM), for three feature vector extraction techniques like GLSDM, SURF filter, and Gabor filter, respectively. The comparisons portrayed in Tables 14.1–14.4 have done with classification efficiency, true positive rate, false positive rate, area under the curve, precision, and F-measure. Different performance matrix can be calculated as:

TABLE 14.1

True positive rate (TPR) & False positive rate(FPR)

	TPR			FPR		
Classifier	GLSDM	Gabor	Surf	GLSDM	Gabor	Surf
Naïve Bayes	0.7	0.74	0.78	0.3	0.18	0.09
KNN Classifier	0.72	0.61	0.73	0.27	0.28	0.15
SVM Classifier	0.72	0.8	0.77	0.25	0.2	0.11
ELM Classifier	0.76	0.82	0.84	0.05	0.11	0.06

TABLE 14.2

Precision & F- measure

	Precision			F-measure		
Classifier	GLSDM	Gabor	Surf	GLSDM	Gabor	Surf
Naïve Bayes	0.73	0.84	0.84	0.68	0.37	0.6
KNN Classifeir	0.75	0.79	0.8	0.75	0.57	0.72
SVM Classifier	0.74	0.8	0.88	0.72	0.8	0.81
ELM Classifier	0.85	0.9	0.91	0.82	0.8	0.88

TABLE 14.3

Area under the curve

	Area Under the Curve		
Classifier	GLSDM	Gabor	Surf
Naïve Bayes	0.66	0.86	0.85
KNN Classifier	0.84	0.82	0.84
SVM Classifier	0.77	0.87	0.88
ELM Classifier	0.94	0.92	0.93

TABLE 14.4

Comparison of efficiencies

	GLSDM		Gabor		Surf	
	Efficiency (%)		Efficiency (%)		Efficiency (%)	
Classifier	Training	Testing	Training	Testing	Training	Testing
Naïve Bayes	75.25	70	85.39	74.19	86.22	76.23
KNN	76.59	72.5	77.52	61.29	78.65	80.22
SVM	81	72.5	100	80.64	100	81.74
ELM	97	90	100	92.5	100	95.03

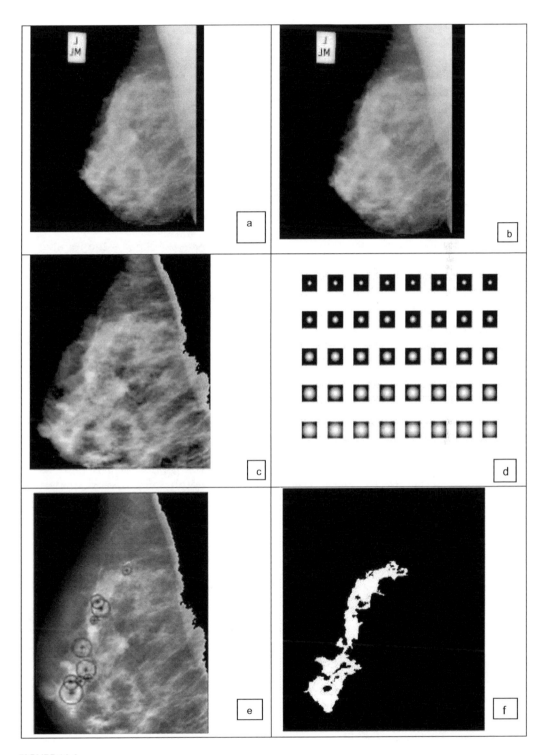

FIGURE 14.4
(a) Input image, (b) denoised image, (c) pectoral muscle removed image, (d) magnitude of Gabor filter features, (e) SURF features, and (f) microcalcification affected area.

$$TPR(\text{True positive rate}) = \frac{TP}{P}$$

$$FPR(\text{False positive rate}) = \frac{FP}{N}$$

$$Pr(\text{Precision}) = \frac{TP}{TP + FP}$$

$$Ac(\text{Accuracy}) = \frac{TP + TN}{P + N}$$

$$Fm(\text{F measure}) = \frac{2}{\frac{1}{Pr} + \frac{1}{TPR}}$$

Here, the term FP denotes the number of false positives, TP denotes the number of true positives, P is the total positive count, and N is the total negative count. The classifier performance is said to be better if it provides high values for TPR, precision, AUC, F-measure, and low values for FPR. All calculations are normalized to the scale of 0 and 1. The mid value 0.5 is chosen for probability cutoff for calculating FPR and TPR.

Table 14.1 shows that extreme learning machine classifier gives good TPR response and lower FPR response for all the feature extraction methods like GLSDM, Gabor, and SURF. It is also observed that SURF outperforms all other feature extraction methods. Table 14.2 depicts that the values of precision and F-measure are better for ELM compared to other classification algorithms.

Table 14.3 shows that AUC is good with ELM. From Table 14.4, it is evident that the mix of ELM and SURF features outperformed the automatic detection of microcalcifications in breast images. This ELM approach is performed by inputting a breast X-ray image from MIAS database and performing necessary morphological operations to remove unnecessary artifacts. The different stages of output of proposed model are shown in Figure 14.4.

14.7 Conclusion

The presented ELM method by using SURF features enhances the efficiency of computer-based microcalcification identification in mammogram images. The label and background are removed by segmenting of the input images, which directs the resizing of the breast images to the exact area of the breast skin line. Both normal and abnormal images and region of interests are traced with the resized image of 32 × 32. SURF surface based features extracted from subimages are taken for calcification identification and the ELM is predicted to be the unparalleled target classifier for this. The single hidden layer based ELM uses defined parameters so that the result is almost accurate as possible (95%). The performance evaluation of ELM with SURF features is carried out by comparing its efficiency and precision and other parameters with other standard classifier and their performance with other feature extraction methods like Gabor and GLSDM. The SURF feature proved as the best out of all other feature vectors. This research will enhance further improvements in works with multiscale features that use SURF and Gabor concurrently which will result in higher efficiencies and lower false positives.

References

[1] M. Bazzocchi, F. Mazzarella, C. Del Frate, F. Girometti, C. Zuiani. CAD systems for mammography: a real opportunity? A review of the literature. Radiol Med 2007 Apr;112(3):329–353.

[2] R. Ramani, N. Suthanthira Vanitha, S. Valarmathy. The pre-processing techniques for breast cancer detection in mammography images. I.J. Image Graph Signal Process 2013;5:47–54.

[3] R. Rouhi, M. Jafari, S. Kasaei, P. Keshavarzian. Benign and malignant breast tumors classification based on region growing and CNN segmentation. February 2015.

[4] T. Huang, G. Yang, G. Tang. A fast two-dimensional median filtering algorithm. IEEE Trans Acoust Speech Signal Process 1979;27:13–18.

[5] A. Sahakyan, H. Sarukhanyan. Segmentation of the breast region in digital mammograms and detection of masses. (IJACSA) Int J Adv Comput Sci Appl 2012;3(2):102–105.

[6] N. Karahaliou Anna, S. Boniatis Ioannis, G. Skiadopoulos Spyros, N. Sakellaropoulos Filippos, S. Arikidis Nikolaos, A. Likaki Eleni et al. Breast cancer diagnosis: analyzing texture of tissue surrounding microcalcifications. IEEE Trans Inform Technol Biomed 2008;12(6):731–738.

[7] M.A. Kupinski, M.L. Giger. Automated seeded lesion segmentation on digital mammograms. IEEE Trans Med Imaging 1998;17(4):510–517.

[8] B. Sahiner, H.-P. Chan, N. Pretrick, D. Wei, M.A. Helvie, D.D. Adler et al. Classification of mass and normal breast tissue – a convolution neural network classifier with spatial domain and texture images. IEEE Trans Med Imaging 1996;15(5):598–609.

[9] H. Eltonsy Nevine, D. Tourassi Georgia, S. Elmaghraby Adel. A concentric morphology model for the detection of masses in mammography. IEEE Trans Med Imaging 2007; 26(6):880–889.

[10] B. Zheng, W. Qian, L.P. Clarke. Digital mammography: Mixed feature neural network with spectral entropy decision for detection of microcalcifications. IEEE Trans Med Imaging 1996;15(5):589–597.

[11] P. Valarmathi, V. Radhakrishnan. Tumor prediction in mammogram using neural network. Glob J Comput Sci Technol 2013;13(2):19–24.

[12] P. Angayarkanni, B. Nadira Banu Kamal. Automatic classification of mammogram MRI using Dendrogram. Asian J Comput Sci Inform Technol 2012;2(4):78–81.

[13] M. Kass, A. Witkin, D. Terzopoulos. Snakes: active contour models. Int J Comput Vis 1988 Jan;1(4):321–331.

[14] T. Chan, L. Vese. Active contours without edges. IEEE Trans Image Process 2001 Feb; 10(2):266–277.

[15] J. Tang, X. Liu. Classification of breast mass in mammography with an improved level set segmentation by combining morphological features and texture features. In Ayman S. El-Baz, Rajendra Acharya U, Andrew Laine and Jasjit S. Suri (ed.) Multi Modality State-of-the-Art Medical Image Segmentation and Registration Methodologies. New York: Springer-Verlag, 2011, pp. 119–135.

[16] C. Li, C.Y. Kao, J.C. Gore, Z. Ding. Minimization of region scalable fitting energy for image segmentation. IEEE Trans Image Process 2008 Oct;17(10):1940–1949.

[17] J.C. Bezdek, R. Ehrlich, W. Full. FCM: the fuzzy c-means clustering algorithm. Comput Geosci 1984;10(2/3):191–203.

[18] K.S. Chuang, H.L. Tzeng, S. Chen, J. Wu, T.J. Chen. Fuzzy c-means clustering with spatial information for image segmentation. Comput Med Imag Graph 2006 Jan;30(1):9–15.

[19] J. Tian, Q. Hu, X. Ma, M. Ha. An improved KPCA/GA-SVM classification model for plant leaf disease recognition. J Comput Inform Syst 2012;18:7737–7745.

[20] M. Melanie. An Introduction to Genetic Algorithms A Bradford Book. The MIT Press, 1999.

[21] L. Crosby Jack. Computer Simulation in Genetics. London: John Wiley & Sons, USA, 1973. ISBN 0-471-18880-8.

[22] A. Ziarati. A multilevel evolutionary algorithm for optimizing numerical functions. IJIEC 2011;2:419430.

[23] H. John. Adaptation in Natural and Artificial Systems. Cambridge, MA: MIT Press, 1992. ISBN 978-0262581110.

[24] O. Babatunde, L. Armstrong, J. Leng, D. Diepeveen. Zernike moments and genetic algorithm: tutorial and application. Br J Math Comput Sci 2014;4(15):2217–2236.

[25] S.N. Sivanandam, S.N. Deepa. Introduction to Genetic Algorithms Springer-Verlag. Heidelberg: Berlin, 2008.

[26] E. Malar, A. Kandaswamy, D. Chakravarthy, A. Giri Dharan. A novel approach for detection and classification of mammographic microcalcifications using wavelet analysis and extreme learning machine. Comput Biol Med 2012;42(9):898–905.

15

Deep Learning Segmentation Techniques for Checking the Anomalies of White Matter Hyperintensities in Alzheimer's Patients

Antonitta Eileen Pious and Sridevi Unni

Department of Computer Applications, Sri Krishna College of Engineering and Technology, Coimbatore, Tamil Nadu, India

15.1 Introduction

The theme of medical image segmentation is to study the anatomical structure; identify the region of interest, that is, lesions and abnormalities; measure the growth of diseases; and helps in treatment planning. The diseased area needs to be segmented and separated from the rest for better understanding and analysis. Segmentation refers to the labeling of pixels into different regions. Segmentation of nontrivial images is one of the most difficult tasks in image processing [3]. Quantitative analysis of brain MRI is routine for many neurological diseases and conditions and relies on accurate segmentation of structures of interest. Deep learning (DL)-based segmentation approaches for brain MRI are gaining interest due to their self-learning and generalization ability over large amounts of data. As the deep learning approaches are becoming more mature, they gradually outperform previous state-of-the-art classical machine learning (ML) algorithms [4]. The disadvantage of using classical ML algorithms was the need to create suitable and specific imaging features that enable further segmentation, which do not generalize well. They need lot of domain expertise, human intervention only capable of what they're designed for; nothing more, nothing less. For AI designers and the rest of the world, that's where deep learning holds a bit more promise. White matter hyperintensities (WMH) are considered to be one of the major signs of small vessel disease on MRI. Damage to small blood vessels in the brain could be the second risk factor leading to Alzheimer's, whereas accumulation of beta amyloid plaques being the first. After a study though, researchers found that both factors were independent predictors of Alzheimer's disease.

15.2 Deep Learning Methodologies for Segmentation of Brain Anomalies

An emerging ML technique is referred to as deep learning. The advantages of DL over classical ML is the self-learning of features which enable in segmentation and creating useful imaging features. When there is lack of domain understanding for feature introspection,

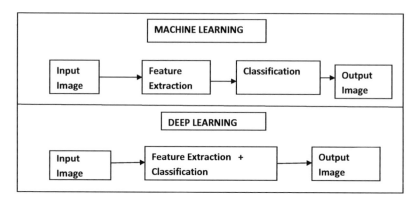

FIGURE 15.1
Example of the learning techniques in ML and DL.

deep learning techniques outshine others as you have to worry less about feature engineering. Shown in Figure 15.1 is the example of learning techniques in ML and DL.

There are several types of deep learning approaches that have been developed for different purposes, such as object detection and segmentation in images. Some of the known deep learning algorithms are deep Boltzmann machines stacked auto-encoders, deep neural networks, and convolutional neural networks (CNNs). Manual segmentation of WM lesions is a time-consuming process. CNNs are the most commonly applied to image segmentation and classification.

15.3 Network Architecture Details

In the network architecture, we can see how CNNs can be used as patch wise, semantic wise, and in cascaded architecture. Fully automated algorithms for WMH lesion detection and segmentation have been the focus of research for many years.

15.3.1 Patch-Wise Segmentation

One of the popular initial deep learning approaches was patch classification where each pixel was separately classified into classes using a patch of image around it. Main reason to use patches was that classification networks usually have full connected layers and therefore required fixed size images. This "patch-wise sampling" ensures that the input has enough variance and is a valid representation of the training dataset. Figure 15.2 shows a patch-wise convolutional neural network (patch-CNN) [5].

Patch-wise convolutional neural network (patch-CNN) refers to a CNN network where patches (i.e., small subset of the entire image) are used for the training process. In this particular study of WMH, a great challenge always faced in WMH segmentation is the inaccuracy of machine learning algorithms in detecting early stages of brain pathology. WMH at early stages are difficult to assess for two main reasons. One is their intricacy, which makes WMH hard to identify even by human eyes and easily mistaken as imaging

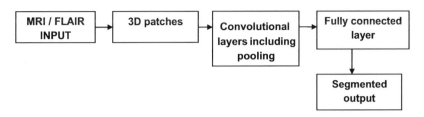

FIGURE 15.2
Patch-wise convolutional neural network (patch-CNN).

artifacts [6]. Another is their small volume. These two facts make the development of automatic WMH segmentations for brains with mild or no vascular pathology challenging [7]. From the study of dice similarity coefficient (DSC) scores, patch-CNN has an output score of 0.80. A good learning algorithm for automatic WMH segmentation should have a good balance in the following different values: DSC, sensitivity (Sen.), specificity (Spe.), and precision (Pre.) scores. From this evaluation of patch-CNN with other classical machine learning algorithms, patch-CNN has outperformed it all. Table 15.1 shows the different values to check the efficiency of a patch-CNN.

In this study (Figure 15.2), a 2D CNN has been used with five convolution layers and two fully connected layers for the purpose of segmentation.

15.3.2 Semantic-Wise Segmentation

In 2014, fully convolutional networks (FCN) by Long et al. from Berkeley popularized CNN architectures for dense predictions without any fully connected layers. This allowed segmentation maps to be generated for image of any size and was also much faster compared to the patch classification approach. Almost all the subsequent state-of-the-art approaches on semantic segmentation adopted this paradigm. Segmentation of brain tissues and WMH of presumed vascular origin is widely being performed in MR images of older patients and is especially relevant in the context of neurovascular and neurodegenerative diseases. In patients suffering from Alzheimer's disease (AD), higher load of WMH has been associated with higher amyloid beta deposits, presence of markers of small vessel disease, and reduced amyloid beta clearance, all these contributing to an overall worsening of the cognitive functions on these patients. Semantic segmentation faces an inherent tension between semantics and location: Global information resolves what while local information resolves where. What can be done to navigate this spectrum from location to semantics? How can local decisions respect global structure? It is not immediately clear that deep networks for image classification yield representations sufficient for accurate, pixel-wise recognition [8]. The spatial output maps of these convolutionalized models make them a natural choice for dense problems

TABLE 15.1

Different values to check the efficiency of a patch-CNN

Method	Threshold	DSC	Sen.	Spe.	Pre.
Patch-CNN	0.81	**0.5376**	**0.9983**	0.5385	0.9974

like semantic segmentation. Similar to auto-encoders, they include encoder part that extracts features and decoder part that upsamples or deconvolves the higher level features from the encoder part and combines lower level features from the encoder part to classify pixels. The input image is mapped to the segmentation labels in a way that minimizes a loss function. Figure 15.3 shows the block diagram for a semantic-wise segmentation.

15.3.3 Cascaded-Wise Segmentation

Unless there is a vast amount of training data the use of CNNs can be a negative impact on the performance of the model on new data. In additional training, such networks may require a long time as well as specific fine-tuning steps. So we should be able to use CNNs in an iterative approach. A simple solution is to train a second CNN which learns to reconstruct from the output of the first CNN. The first CNN could be trained to detect a region of WMH which could be used as an attention model, rather than dealing with the entire information, the second CNN could be using the distilled information from the first CNN. In fact, we can concatenate a new CNN on the output of the previous CNN to build extremely deep networks which iterate between intermediate de-aliasing and the data consistency reconstruction .We term this as cascading network. This kind of network is very efficient as it avoids the influence of surrounding noise. Figure 15.4 shows us the block diagram for a cascaded-wise segmentation.

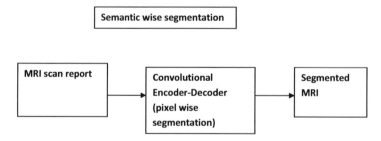

FIGURE 15.3
Semantic-wise segmentation.

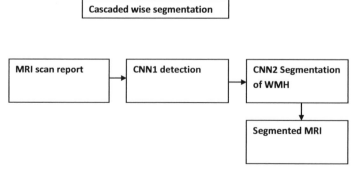

FIGURE 15.4
Cascaded-wise segmentation.

15.4 Summary

Despite the significant effort in brain lesion segmentation and advanced imaging techniques, accurate segmentation of brain lesions remains a challenge. Many automated methods have been proposed for lesion segmentation problem, including unsupervised modeling methods that aim to automatically adapt to new image data [9–11]. Despite the significant impact of deep learning techniques in quantitative brain MRI, it is still challenging to have a generic method that will be robust to all variations in brain MR images from different institutions and MRI scanners. The performance of the deep learning methods depends highly on several key steps such as preprocessing, initialization, and postprocessing.

References

[1] www.nia.nih.gov/health/what-happens-brain-alzheimers-disease

[2] www.ninds.nih.gov/Disorders/Patient-Caregiver-Education/Life-and-Death-Neuron

[3] Anubha et al., A Review on MRI Image Segmentation Techniques, International Journal of Advanced Research in Electronics and Communication Engineering (IJARECE), Volume 4, Issue 5, May 2015.

[4] Zeynettin Akkus et al., Deep Learning for Brain MRI Segmentation: State of the Art and Future Directions, Journal of Digital Imaging, Volume 30, Issue 4, August 2017.

[5] Muhammad Febrian Rachmadi et al., Deep Learning vs. Conventional Machine Learning: Pilot Study of WMH Segmentation in Brain MRI with Absence or Mild Vascular Pathology, Journal of Imaging, Dec 2017.

[6] Muhammad Febrian Rachmadi, Maria del C. Valdés-Hernández, Maria Leonora Fatimah Agan, Taku Komura. Evaluation of Four Supervised Learning Schemes in White Matter.

16

Investigations on Stabilization and Compression of Medical Videos

D. Raveena Judie Dolly and D. J. Jagannath

Electronics and Communication Engineering, Karunya Institute of Technology and Sciences, Coimbatore, Tamil Nadu, India

R. Anup Raveen Jaison

SASTRA Deemed University, Thanjavur, Tamil Nadu, India

16.1 Introduction: Background and Objective

Clinical videos with jittery platform require stabilization to ascertain stability. Differential motion estimation can be adopted if handheld cameras are utilized. After the entire video is converted to frames, filtering algorithm can be adopted to undergo corrective measures in the motion as suggested in ref. [1]. Numerous research has been evolved in the area of video compression and video stabilization. Video compression involving adaptive frame determination (AFD) resulted in better subjective and objective result as provided in ref. [2]. After performing the adaptive frame determination, it is exposed to affine translation in order to reduce the buffering of memory. After which, the whole system undergoes affine transformation as suggested in ref. [3]. The information is stored in terms of parameters in a matrix fashion which occupies very less file size. This is fed to an optimizer to get optimized results as indicated in ref. [4] so as to get very good results. Both Broyden–Fletcher–Goldfarb–Shanno (BFGS) and limited-memory BFGS (L-BFGS) belong to the family of quasi-Newton methods. L-BFGS is preferred in many applications so as to reduce computer memory utilization.

16.2 Investigations on Stabilization and Compression

Stabilization is performed in order to reduce unwanted camera disturbances and jittery movements. The stabilization undergoes a process of computation of affine parameters. Smoothing is performed followed by warping process. Motion estimation is performed by adopting the techniques followed in refs. [5–7]. Compression is followed after stabilizing the video as shown in Figure 16.1.

Once performing compression, to further enhance the quality certain optimization techniques are performed to the medically compressed video. Performance metrics like PSNR and SSIM are calculated to ensure better quality.

FIGURE 16.1
Overall block view.

16.3 Implementation and Results

MRI slices from dicom are read for 37 recordings. The case study was analyzed with the age group from 19 to 62. Few outcomes are analyzed in detail and are projected in the table. Patient code is taken as PC. Adaptive frame determination was performed where the threshold was fixed as 0.8. The outcome obtained was without degradation. To further enhance the quality, few optimizers were chosen and executed in MATLAB.

The obtained results as shown in Table 1 indicated in ref. [2] along with the proposed method were graphically obtained for PSNR and SSIM.

It was observed in Figure 16.2 that the implementation with L-BFGS optimizer provides better results than the BFGS optimizer. The BFGS optimizer gives better results than the implementation of AFD.

Comparatively, PSNR of BFGS optimizer provides a substantial increase of +2.73 dB with AFD with I and P frames and a +2.33 dB increase with L-BFGS optimizer in comparison with BFGS optimizer.

TABLE 1

Implementation with AFD and optimizers

Cases	Implementation of AFD with I and P frames [2]		Implementation with BFGS optimizer		Implementation with L-BFGS optimizer	
	PSNR	**SSIM**	**PSNR**	**SSIM**	**PSNR**	**SSIM**
MRI brain-PC:2K17006	36.53	0.88	36.8	0.89	37.1	0.9
MRI brain-PC:2K17027	36.01	0.82	36.41	0.84	36.87	0.86
MRI brain-PC:2K17032	35.98	0.83	36	0.84	36.24	0.85
MRI brain-PC:2K17057	36.21	0.8	36.78	0.82	36.92	0.83
MRI brain-PC:2K17106	36.19	0.87	36.52	0.88	36.8	0.89
MRI brain-PC:2K17121	36.25	0.88	36.67	0.89	36.81	0.9
MRI brain-PC:2K17143	36.67	0.86	36.90	0.88	37	0.9
MRI brain-PC:2K17161	35.91	0.82	36.09	0.84	36.34	0.86
MRI brain-PC:2K17183	35.77	0.83	36.02	0.85	36.32	0.86
MRI brain-PC:2K17194	35.93	0.84	35.99	0.86	36.11	0.88

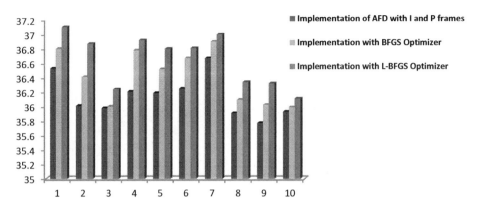

FIGURE 16.2
PSNR of 10 different cases.

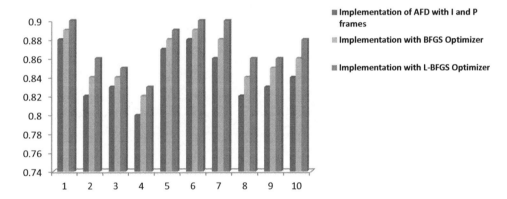

FIGURE 16.3
SSIM of 10 different cases.

It can also be observed in Figure 16.3 an increase of SSIM as 0.16 while adopting BFGS optimizer in comparison with AFD with I and P frames and an increase of 0.14 with L-BFGS optimizer in comparison with BFGS optimizer.

The results and observations clearly indicate that the objective evaluation with PSNR and SSIM has a substantial increase in L-BFGS optimizer.

16.4 Conclusion

Few medical huge sized videos are identified for performing compression. It was observed that those videos had a jittery platform. In order to ensure better quality in output video after performing compression, stabilization is performed before compression. To ensure good quality, certain optimization techniques are performed. It was observed that after using suitable optimizers, an increase in PSNR and SSIM was observed.

References

[1] Marius Tico and Markku Vehvilainen. 2007. Robust Method of Video Stabilization. 15th European Signal Processing Conference (EUSIPCO 2007), Poznan, Poland, September 3–7, 2007.

[2] Raveena Judie Dolly D., Jagannath D.J., Dinesh Peter J. (2019) Video Stabilization for High-Quality Medical Video Compression. In: Peter J., Fernandes S., Eduardo Thomaz C., Viriri S. (eds) Computer Aided Intervention and Diagnostics in Clinical and Medical Images. Lecture Notes in Computational Vision and Biomechanics, vol. 31. Springer, Cham.

[3] Dolly, D.R.J. , Bala, G.J. , Peter, J.D. 2017. Performance Enhanced Spatial Video Compression Using Global Affine Frame Reconstruction. Journal of Computational Science, Vol. 18, pp. 1–11.

[4] Dolly, D.R.J. , Bala, G.J. , Peter, J.D. 2018. A Hybrid Tactic Model Intended for Video Compression Using Global Affine Motion and Local Free-Form Transformation Parameters. Arabian Journal for Science and Engineering, Vol. 43, No. 8, pp. 1–15.

[5] Paresh Rawat and Jyoti Singhai. 2011. Review of Motion Estimation and Video Stabilization Techniques for Hand Held Mobile Video. Signal & Image Processing: An International Journal (SIPIJ), Vol. 2, No. 2, pp. 159–168.

[6] Karra Somasekhar Reddy, Karra Somasekhar Reddy, and S. Akhila. 2014. A Survey on Video Stabilization Algorithms. International Journal of Advanced Information Science and Technology (IJAIST), Vol. 31, No. 31. ISSN: 2319: 2682, pp. 1–20.

[7] S.M. Kulkarni, D.S. Bormane, and S.L. Nalbalwar. 2015. Coding of Video Sequences Using Three Step Search Algorithms. Procedia computer science, vol. 49, pp. 42–49.

17

An Automated Hybrid Methodology Using Firefly Based Fuzzy Clustering for Demarcation of Tissue and Tumor Region in Magnetic Resonance Brain Images

Saravanan Alagarsamy, Kartheeban Kamatchi, and Vishnuvarthanan Govindaraj
Kalasalingam Academy of Research and Education, Krishnankoil, Tamil Nadu, India

17.1 Introduction

Cancer cells are responsible for developing tumor in the human body. The grade of tumors present in the human brain can be classified into two types: (i) Benign (low grade tumor) this type of tumor are less aggressive and can be cured if the patient is monitored continuously by radiologist and (ii) Malignant (high grade tumor) types are more aggressive and it can be diagnosed in the early stage [1].

The malignant tumors are diagnosed by computerized tomography (CT), positron emission tomography (PET), and magnetic resonance imaging (MRI). Different types of image sequence can be done for patients to obtain the required information for decision making. MRI tends to be a more effective technique for early diagnosis of tumor [2]. But brain tumor with different grades and volume remain a challenge for MRI diagnosis. For particular cases, MRI technique is not sufficient for decision making and some manual assistance is required for accurate brain tumor segmentation. Demarcation of brain tumor is a very difficult task for analyzing the medical images. There is a need of an automated segmentation technique for identifying the tumor and tissue regions present in MR brain image. Such problem can be effectively handled with the help of recommended Firefly based IT2FCM [3].

MRI scanner produces different image sequences of human brain like T1-weighted (T1-W), T2-weighted (T2-W), fluid attenuated inversion recovery (FLAIR), and multi planar reconstruction (MPR) with contrast enhancement. Different axes such as axial, coronal, and sagittal are effectively segmented by the proposed Firefly based IT2FCM algorithm. This process enables the radiologist to have a better view of heterogeneous tumor types resulting in better diagnosis of patients [4].

17.2 Related Works

By analyzing the related research works carried for MR brain image segmentation, an automated hybrid technique to perform tumor identification and better visualization of

TABLE 17.1

Overview of the conventional works done so far in medical image processing

S. No.	Name of the contributors	Segmentation techniques used by the researchers	Pros of the technique	Cons of the technique
1.	Pinto et al. [5]	The authors developed hierarchical technique for brain tumor segmentation using randomized tree search.	This technique segment T1-weighted and T2-weighted in MRI image.	Dice Similarity coefficient index (DOI) obtained by the technique is 0.85,which can be further extended.
2.	Selvapandian et al. [6]	The author formulated the fuzzy Inference System based Adaptive Neuro (ANFIS) for segmentation.	Low grade and high grade gliomas in MR images can be demarcated by the suggested technique.	The proposed method achieves 91.9% of sensitivity, which can be further enhanced.
3.	Gooya et al. [7]	The authors used segmentation and registration technique for gliomas called as GLISTR.	The proposed technique was used to segment the gliomas present in different sequences of MR images.	The high grade gliomas located in the human brain are detected perfectly, but the segmentation process can be applied toward low grade gliomas.
4.	Raju et al. [8]	The authors used Harmony-Crow Search (HCS) Optimization technique for tumor segmentation.	The Harmony-Crow Search (HCS) Optimization technique identifies the tumor and edema portion in MRI.	The suggested technique produces better results for high grade tumor only.
5.	Ma et al. [9]	The authors used concatenated and connected random forests (CCRF) method for segmentation.	The authors proposed an automated and accurate brain tumor segmentation for multimodal MRI	The Dice score produced by the proposed technique is 0.84, which can be further improved for accurate decision making.
6.	Singh and Bala. [10]	The authors proposed a domain approach using the combination of discrete cosine transform (DCT) and fuzzy C-means (FCM) techniques for the demarcation of tissues in MR brain images.	Demarcation of GM and WM in T2-weighted brain images are perfectly done by the suggested technique.	The authors have mainly focused on T1-weighted images. It can be further extended toward T2-weighted images.
7.	Jothi and Inbarani. [11]	The authors used the features of Rough Set Firefly based Quick Reduct (TRSF-FQR) for segmentation.	The tumor region located in MR brain slices are identified by the suggested methodology.	The proposed methodology produced good segmentation results but has been tested for a limited set of datasets.
8.	Tong et al. [12]	Combination of features sparse kernel coding texture features were used by the authors for the delineation of FLAIR images.	Fluid attenuated inversion recovery (FLAIR) sequences are precisely segmented by the automated technique.	The proposed segmentation technique concentrates much on single-modal data sources and the sensitivity produced was 0.84. The sensitivity value obtained can be increased.
9.	Lim and Mandava et al. [13]	The authors applied random walks based on homogeneity and object feature called as HORW for segmentation.	Multi sequence brain tumors present in MRI were segmented by the authors' proposed semi-automatic approach.	The DOI produced is 0.7 for high grade and 0.63 for low grade tumor and can be further increased.

(Continued)

TABLE 17.1 (Cont.)

S. No.	Name of the contributors	Segmentation techniques used by the researchers	Pros of the technique	Cons of the technique
10.	Angulakshmi and Priya [14]	The authors used the techniques spectral clustering with region of interest (ROI) for demarcating tumor and tissue portions in MRI.	The edema and tumor portions are clearly segmented by the suggested technique.	The suggested technique exemplified a DOI value of 0.7 for edema region that needs to be improved.
11.	Soltaninejad et al. [15]	The authors utilized learning method dependent upon super voxel for tumor segmentation and the multi sequence MRI images.	The algorithm provides good segmentation and delineation across all grades of tumor present in MR brain image.	The automatic tumor segmentation technique produces a Dice score of 0.84 against ground truth can be further extended.
12.	Kamathe and Joshi [16]	Novel automated tissue segmentation technique based on band expansion process (BEP) combination of independent component analysis (ICA) was used by the authors.	The suggested technique can demarcate the tissue in brain such as cerebro spinal fluid (CSF), gray matter (GM), and white matter (WM).	The authors mainly focused on T1-W and T2-W images. It can be further applied to FLAIR images.
13.	Pham et al. [17]	Combination of particle swarm optimization (PSO) and clustering is done by kernelized fuzzy entropy, local spatial information is used for grouping with integration of bias correction.	Identification of tumor and segmentation done for T1-W MRI images.	Segmentation performance of the techniques may decrease while high intensity of noise along with intensity and non-uniformity (INU) object are included in the MRI data.
14.	Li et al. [18]	The authors used maximum a posteriori probabilistic (MAP) technique for identifying the brain tumor.	The proposed technique is able to segment the portions of gliomas in the MRI slices.	The proposed technique produced DOI for tumors of high-grade data as 0.73, 0.56, and 0.5, and they can be further increased.

edema region, tissue structures, cerebrospinal fluid (CSF), gray matter (GM), and white matter (WM) is needed, and also this type of method is not yet used for medical analysis. Heterogeneous types of tumor with different grades are identified and tissue segmented with the help of proposed Firefly based IT2FCM technique in less processing time. This algorithm deliberates good result for tumors with complex structures that are located in various parts of the brain.

17.3 Methodology

In the proposed methodology, first, preprocessing of the input image is done (skull stripping). Second, skull stripped image is given as input to the Firefly clustering algorithm. Third, augment the current best solution and update the membership function. This process is repeated to find the best position value, and fourth, acceptance of new solution, and fifth, check out for the best position. Finally, clustering is done by IT2FCM.

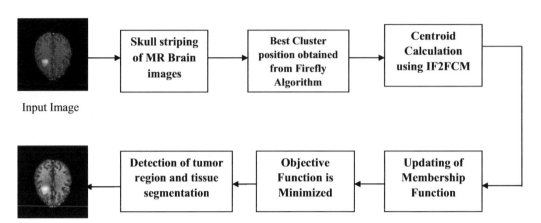

Input Image

Segmented Image

FIGURE 17.1
Block diagram of brain tumor segmentation based on Firefly based IT2FCM.

17.3.1 Firefly Algorithm

Firefly is an optimization technique inspired by the biological characteristics of a firefly, and it is a meta-heuristic procedure used to optimize a problem. In the proposed work, an algorithm inspired by the natural phenomenon of firefly has been used to solve the optimization problem [19]. Firefly is a winged beetle found commonly in the country side. These flies are especially known for their brilliant use of bioluminescence during twilight to attract its mates or prey. The meta-heuristic algorithm is mainly used to enhance the performance for segmentation process.

The Firefly algorithm is developed with the following conditions:

(i) All fireflies are unisexual and are attracted to each other regardless of the gender.

(ii) The luminance value is proportional to attractiveness between two fireflies.

(iii) The fitness or cost function is measured by the luminance of a firefly and is presented in the analytical form that is used to help the search process. For example, in optimization problem, the value of the cost is directly proportional to the luminance of the fitness function [20].

Step 1: (Generate the initial population of firefly) The Fireflies generates N solutions, which are randomly distributed and it is assumed to be the initial population ($i = 1, 2, ..., N$).

Step 2: (Measuring and finding the new solution) The position update can be calculated between the movements of the attracted firefly i toward a brighter firefly j, and can be defined in the following equation:

$$v_i^{t+1} = v_i^t + \beta 0 e^{-\gamma d^2} ij\left(v_j^t - v_i^t\right) + rand_1 \qquad (17.3.1)$$

where v_i^t - is the new updated position of firefly
$v_i^{(t+1)}$ - is the initial/starting position of firefly
$\beta 0 e^{-\gamma d^2 ij}\left(v_j^t - v_i^t\right)$ is assumed as attractiveness force between two fireflies.

Step 3: (Augmenting the current best solution) Every solution of v_i^t is assigned with the updated position v_i^{t+1} where uniform random number ($rand_1$) lies in the range between 0 and 1.

Step 4: (Acceptation of a new solution) Measure the set of solutions to obtain the updated fitness value. After each step, the new/updated fitness value is calculated and then compared with the old/fitness value. If you find that obtained new fitness value is better than the old fitness value, modify the best position corresponding to the new fitness value.

Step 5: (Check the stopping criteria) Iterate the process until the best fitness and best position is identified. Then the process will be stopped.

17.3.2 Interval Type-2 Fuzzy C-Means

The logic of fuzzy set produces the optimized solution for uncertainty problem. The fuzzy logic of type-2 is updated of the concept of fuzzy logic of type-1. The reason for selecting the fuzzy logic of type-2 is used for providing improved solution for the optimization problem [21].

$$\tilde{A} = \left\{ ((x, u), \mu_{\tilde{A}}(x, u)) | \forall u \in J_x \subseteq [0, , 1] \right\} \tag{17.3.2.1}$$

The membership function $\mu_{\tilde{A}}(x, u)$ range is described as $0 \leq \mu_{\tilde{A}}(x, u) \leq 1$, and the fuzzy set $\tilde{A}$ can be further expressed as:

$$\tilde{A} = \int_{x \in X} \int_{u \in J_x} \frac{\mu_{\tilde{A}}(x, u)}{(x, u)} J_x \subseteq [0, 1] \tag{17.3.2.2}$$

The amalgamation function of all acceptable variables of input x and u is specified by the double integration $\int\int$ in (17.3.2.2). The x primary membership function is defined by the parameter $J_x \subseteq [0, 1]$, and the secondary fuzzy set is denoted as $\mu_{\tilde{A}}(x, u)$. Further, membership function used for type-2 grade can possess a subset of range between 0 and 1.

17.3.3 Proposed Firefly Based IT2FCM

The suggested methodology follows the consecutive methods to produce the segmentation of MR images:

Step 1: Calculation of Euclidean distance $d_{ik}^2 = \|v_k - v_i\|$ indicates the distance among the input data v_k and the cluster center v_i. The Euclidean norms are inferred by the following notation $\| \|$, and total number of cluster denoted by c. The total number of cluster is considered as "3" ($3 \leq c \leq N$).The total count of pixels in the input image is represented as N.

Step 2: Membership functions of upper bound $\bar{\mu}_{\tilde{A}}(x)$ and membership function of lower bound $\underline{\mu}_{\tilde{A}}(x)$ of IT2FLS are utilize to control the uncertainty of fuzzifier "m." Upper and lower membership functions are calculated as mentioned in (17.3.3.1) and (17.3.3.2).

$$\bar{\mu}_{\bar{A}} = \bar{\mu}_{ik} = \begin{cases} \dfrac{1}{\sum_{j-1}^{c} \left(\frac{d_{ik}}{d_{jk}}\right)^{\frac{2}{m_1-1}}}, & \sum_{j=1}^{c}\left(\frac{d_{ik}}{d_{jk}}\right) < c \\[4mm] \dfrac{1}{\sum_{j-1}^{c}\left(\frac{d_{ik}}{d_{jk}}\right)^{\frac{2}{m_2-1}}}, & \text{otherwise} \end{cases} \qquad (17.3.3.1)$$

$$\underline{\mu}_{\bar{A}} = \underline{\mu}_{ik} = \begin{cases} \dfrac{1}{\sum_{j-1}^{c}\left(\frac{d_{ik}}{d_{jk}}\right)^{\frac{2}{m_1-1}}}, & \sum_{j=1}^{c}\left(\frac{d_{ik}}{d_{jk}}\right) \geq c \\[4mm] \dfrac{1}{\sum_{j-1}^{c}\left(\frac{d_{ik}}{d_{jk}}\right)^{\frac{2}{m_2-1}}}, & \text{otherwise} \end{cases} \qquad (17.3.3.2)$$

Step 3: Cluster center (R) is calculated by the incorporation with the optimal solution provided by Firefly algorithm of v_i^{t+1}. The optimal clusters R_L and R_H are formulated using Firefly $\left(v_i^{t+1}\right)_{\text{low}}$ and $\left(v_i^{t+1}\right)_{\text{high}}$:

$$R = \left(\left(v_i^{t+1}\right)_{\text{low}} + \left(v_i^{t+1}\right)_{\text{high}}\right)/2 \qquad (17.3.3.3)$$

Step 4: The membership function μ_{ik} is obtained using the calculation:

$$\mu_{ik} = \frac{\mu_{ik}^L + \mu_{ik}^R}{2} \qquad (17.3.3.4)$$

Step 5: The final segmentation result of Firefly-IT2FCM is provided by the membership function μ_{ik} by examining the limit of upper membership and lower membership functions.

17.4 Experimental Results and Discussion

The datasets containing tumor with high and low grade used in this chapter obtained from BRATS 2013 challenge have ground truth results, and they are examined to understand the accuracy of segmentation of the proposed methodology. The images used in this chapter have been fixed to a uniform size of 256 × 256 (pixel values) so as to realize the segmentation property possessed by the developed Firefly based IT2FCM technique. This work was experimented with the computer provisioned with Pentium processor (7th generation) with a RAM space of 8 GB and operating in a clock speed of 2.6 GHz.

The Firefly based IT2FCM is applied in different modalities of MRI brain slices (T1-Weighted, T2-Weighted, and FLAIR) images. Good segmentation of WM and GM structures and better tumor region identification are conferred by the proposed methodology and its segmentation results can be visually compared with PSO-FCM results mentioned in Figure 17.2. The segmented results of Firefly based IT2FCM show the clear identification of boundaries for tumor and tissue objects for all high grade tumors and low grade tumors present in MR brain images. The tumors present in multiple locations and edema portions were clearly identified by the proposed Firefly based IT2FCM. The

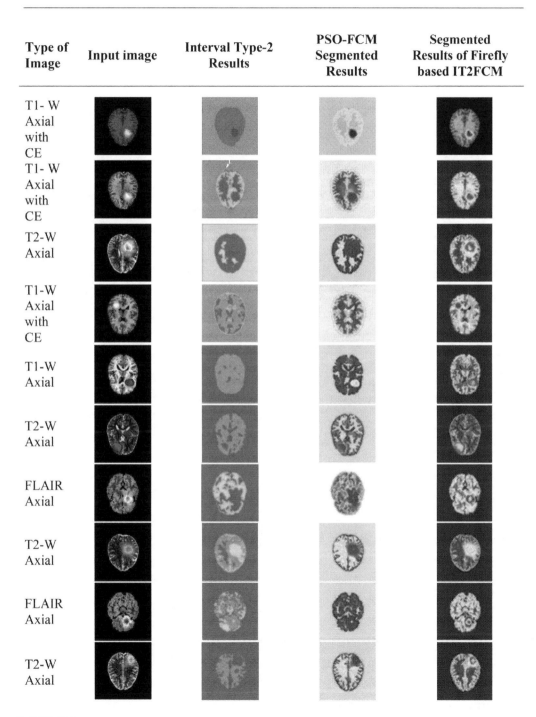

FIGURE 17.2
Identification of high and low grade tumors in MR brain images acquired from BRATS-2013 Challenge database.

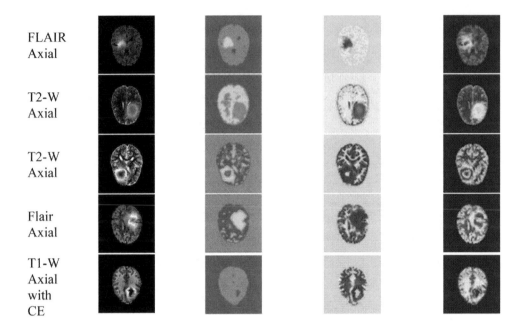

FLAIR
Axial

T2-W
Axial

T2-W
Axial

Flair
Axial

T1-W
Axial
with
CE

FIGURE 17.2
(Cont.)

proposed technique will support the medical image analysis by identifying the tumor with different boundaries to make prediction in a more accurate way.

The segmentation analysis of the proposed technique (for both low grade and high grade tumors) is described in Table 17.2. Various performance metrics are used for the

TABLE 17.2

Performance evaluation of segmented Firefly-IT2FCM algorithm for BRATS-2013 challenge dataset)

S. No	Type and sequences of input image	MSE	PSNR in dB	TC	DOI	Elapsed time in seconds
1.	T1- W Axial with CE	0.0013	76.9913	0.9218	0.9535	2.62785
2.	T1- W Axial with CE	0.0010	78.130	0.9448	0.9789	2.32561
3.	T2-W Axial	0.0015	76.369	0.9143	0.9463	2.826976
4.	T1-W Axial with CE	0.0012	77.3389	0.9398	0.9712	2.758529
5.	T1-W Axial	0.0015	76.369	0.9143	0.9463	2.826976
6.	T2-W Axial	0.0010	78.130	0.9448	0.9789	2.32561
7.	FLAIR Axial	0.0012	77.3389	0.9398	0.9712	2.606782
8.	T2-W Axial	0.0013	76.9913	0.9218	0.9535	2.62785
9.	FLAIR Axial	0.0015	76.369	0.9143	0.9463	2.826976
10.	T2-W Axial	0.0010	78.130	0.9448	0.9789	2.32561
11.	FLAIR Axial	0.0012	77.3389	0.9398	0.9712	2.758529
12.	T2-W Axial	0.0016	76.089	0.8926	0.9326	2.925642
13.	T2-W Axial	0.0010	78.130	0.9448	0.9789	2.32561
14.	Flair Axial	0.0012	77.3389	0.9398	0.9712	2.758529
15.	T1-W Axial with CE	0.0016	76.089	0.8926	0.9326	2.925642

S.NO	Type of Image	Region of Tumor detected using Firefly based IT2FCM	Ground Truth Images of BRATS-2013 Challenge
1	T1- W Axial with CE		
2	T1- W Axial with CE	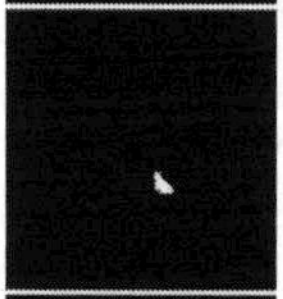	
3	T2-W Axial	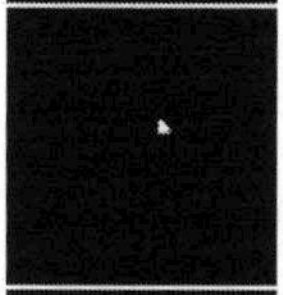	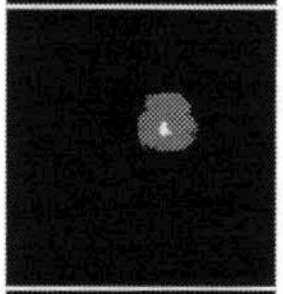
4	T1-W Axial with CE		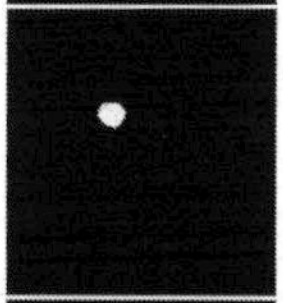
5	T1-W Axial		
6	T2-W Axial		
7	FLAIR Axial	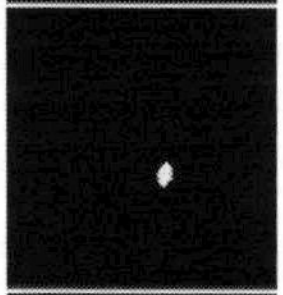	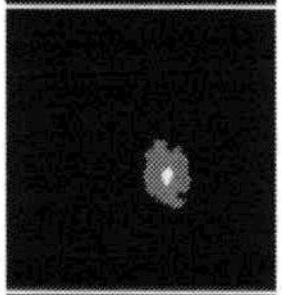
8	T2-W Axial	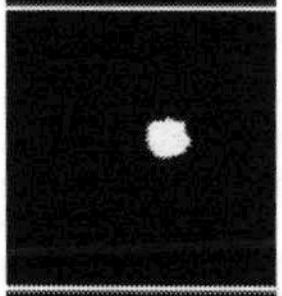	
9	FLAIR Axial	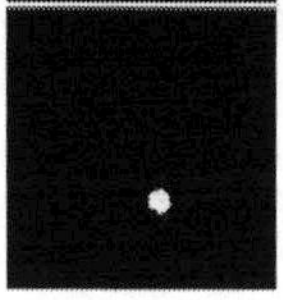	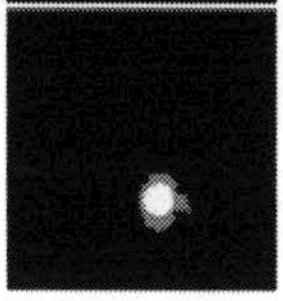

FIGURE 17.3
High and low grade tumors identified by the proposed Firefly based IT2FCM compared with ground truth Images of MR brain images acquired from BRATS-2013 Challenge database.

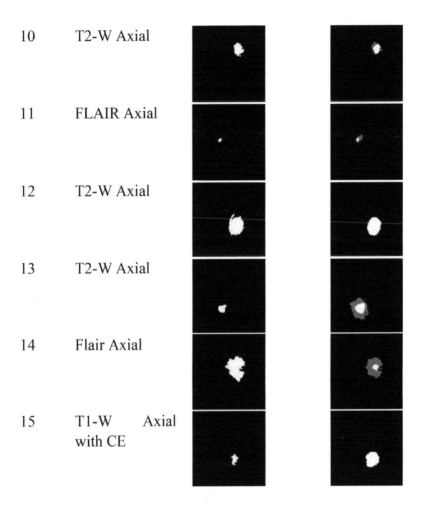

10	T2-W Axial
11	FLAIR Axial
12	T2-W Axial
13	T2-W Axial
14	Flair Axial
15	T1-W Axial with CE

FIGURE 17.3
(Cont.)

validation of recommended Firefly based IT2FCM. The values obtained for each images produced by the proposed technique are shown in Table 17.2.

Identification of tumor portion and demarcation of tissues for high grade and low grade are impressive on comparing it with the ground truth images of BRATS-2013 Challenge database, and this has been proved by visualizing the recommended Firefly based IT2FCM results in Figure 17.3.

17.4.1 Mean Squared Error (MSE)

MSE indicates the squaring of the differentiated magnitude. The squared errors of average are represented, and the values are obtained by deducting the input value from the output value. Mean square error indicated the cumulative error value from input $C(i,j)$ and segmented output $D(i,j)$, and then squared.

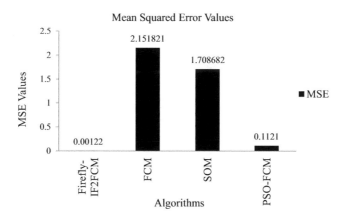

FIGURE 17.4
Performance evaluation of mean squared error (MSE) values.

$$MSE = \frac{1}{mn}\sum_{i=0}^{m-1}\sum_{j=0}^{n-1}[C(i,j) - D(i,j)]^2 \qquad (17.4.1.1)$$

In Equation (17.4.1.1), number of rows and columns are denoted as m and n for the input image. Here, $i = m$ and $j = n$ are used for representing the "rows" and "columns" for incremental operation [22].

The proposed Firefly based IT2FCM produces an average MSE value of 0.001273 and is quite superior to the PSO-FCM and other conventional techniques, and it produces better outcome in identifying tumor and edema regions [23,24].

17.4.2 Peak Signal to Noise Ratio (PSNR)

Interference of noise for segmented slices can be calculated by using the parameter PSNR. The PSNR value obtained by the projected technique is high and it indicates that noise interference signal acting upon the magnetic resonance image is low. The upper limit pixel value for MR image is denoted as 255. The values of PSNR are reliant upon mean square error values. In cognizance, the techniques used to perform segmentation should produce PSNR value between 40 and 100 dB, which could be acceptable in presence of noise. Segmented techniques mainly rely on PSNR value that can be for transmission of accurate results. Depending upon the PSNR value, resistance of noise will be measured for segmenting output image [25,26].

$$PSNR = 10\log_{(10)}((MAX_i^2)/MSE) = 20\log_{(10)}(MAX_i/MSE)$$
$$= 20\log_{10}(MAX_i) - 10\log_{10}MSE(9) \qquad (17.4.2.1)$$

The proposed segmentation algorithm delivered a PSNR value of 77.1428 dB for brain MR slices. Appreciable PSNR values obtained by Firefly based IT2FCM are quite impressive in segmentation and produce better than the conventional algorithms.

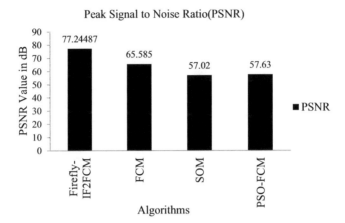

FIGURE 17.5
Performance evaluation of peak signal to noise ratio (PSNR) values in decibels.

17.4.3 Jaccard (Tanimoto Co-efficient) Index

The similarity of pixels between the segmented resulted image (P) and the ground truth image (G) can be expressed by Jaccard index (TC). The TC can be evaluated by union and intersection function of ground truth image (P) and segmented image (S) [13,27].

$$J(G,P) = \frac{S(G \cap P)}{S(G \cup P)} \tag{17.4.3.1}$$

Better segmentation accuracy can be obtained by improving the TC values. TC values are represented as Equation (17.4.3.1).The Firefly based IT2FCM produces an average TC value of 92.734% for segmenting the high grade tumors, which is quite higher than the FCM, SOM, and PSO-FCM. The proposed algorithm renders appreciable segmentation results for high grade tumors.

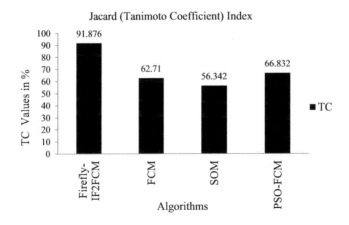

FIGURE 17.6
Performance evaluation of TC values in percentile.

17.4.4 Dice Overlap Index (DOI)

Dice score value is calculated using Jaccard index $J(G, P)$ values [28,29]. Dice score can be measured by finding the overlapping pixels available in the segmented resulted image (P) and ground truth image (G). Higher DOI value indicates better segmentation accuracy for identifying the tumor present in MR brain image. DOI is calculated using Equation (17.4.4.1).

$$D(G, P) = 2 * \frac{J(G, P)}{1 + J(G, P)} \qquad (17.4.4.1)$$

The suggested methodology delivered an average DOI value of 96.07%, which is quite impressive than the conventional techniques. The performance accuracy of the algorithm can be measured by DOI. By producing more DOI values better tumor and tissues can be identified and segmented. Firefly based IT2FCM produces higher DOI value than other comparative techniques. Figure 17.7 clearly shows that the Firefly based IT2FCM produces better segmentation accuracy than the conventional (FCM, SOM, and PSO-FCM) techniques [30].

17.4.5 Computational Time

The amount of time needed to accomplish the segmentation process of the input images is measured in terms of elapsed time, and then usually specified in seconds [31,32]. The developed Firefly based IT2FCM consumes an average 2.6515 seconds for segmenting the high grade and low grade tumors present in the MR brain image. The suggested techniques take less time for performing the task without affecting the efficiency of the segmentation. Some of the techniques produce faster results, but it takes more time for performing the segmentation. But the recommended Firefly based IT2FCM technique produces more accurate result with less time taken for performing the segmentation of input images.

Figure 17.8 shows the time taken for performing the segmentation of the proposed and various conventional techniques [17].

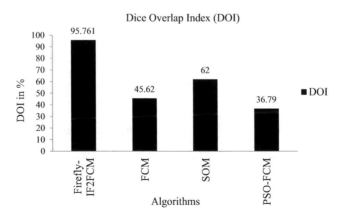

FIGURE 17.7
Performance evaluation of DOI in percentile.

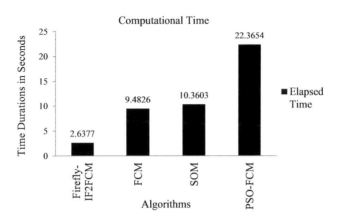

FIGURE 17.8
Evaluating the performance of computational time for proposed techniques.

17.5 Conclusion

An automated hybrid technique (Firefly based IT2FCM) is recommended through this chapter, and the integrated technique requires less time for segmenting different types (T1-Weigted, T2-Weighted, FLAIR) of MR brain slices. The performance of the Firefly technique is improved by adding IT2FCM, which provides better segmentation results for different MR brain slices. The proposed technique identifies the high grade tumor region and segments the tissue structure better than existing techniques such as (FCM, SOM and PSO-FCM). The efficiency of the suggested technique is proved by using comparison parameters such as MSE, PSNR, TC and DOI score values, which is found to be improved when compared to conventional techniques. Appreciable demarcation of brain tissue structures is accomplished using the recommended technique. The suggested technique (Firefly based IT2FCM) will be of much help to the radiologists to detect the pathologies and tumor regions located within different organs of the human body in terms of accurate decision making (especially in identifying the portion of tumor and demarcation of the tissues).

References

[1] Ahmadvand, Daliri. "Improving the runtime of MRF based method for MRI brain segmentation." *Applied Mathematics and Computation* 25 no. 1 (2015): 808–818.

[2] Vishnuvarthanan, Rajasekaran, Subbaraj et al. "An unsupervised learning method with a clustering approach for tumor identification and tissue segmentation in magnetic resonance brain images." *Applied Soft Computing* 38(2016):190–212.

[3] Demirhan, Guler. "Combining stationary wavelet transform and self-organizing maps for brain MR image segmentation." *Engineering Applications of Artificial Intelligence* 24 no. 2 (2011): 358–367.

[4] Abdel-Maksoud, Elmogy, Al-Awadi. "Brain tumor segmentation based on a hybrid clustering technique." *Egyptian Informatics Journal* 16 (2015):71–81

[5] Pinto, Pereira, Rasteiro, Silva. "Hierarchical brain tumour segmentation using extremely randomized trees." *Pattern Recognition* 82 (2018): 105–117.

[6] Selvapandian, Manivannan. "Fusion based glioma brain tumor detection and segmentation using ANFIS classification." *Computer Methods and Programs in Biomedicine* 166 (2018): 33–38.

[7] Gooya, Pohl, Bilello, Cirillo, Biros, Melhem, Davatzikos. "GLISTR: Glioma image segmentation and registration." *IEEE Transactions on Medical Imaging* 31 no. 10 (2012): 1941–1954.

[8] Raju and Suresh, Rao. "Bayesian HCS-based multi-SVNN: A classification approach for brain tumor segmentation and classification using Bayesian fuzzy clustering." *Biocybernetics and Biomedical Engineering* 38 no. 3 (2018): 646–660.

[9] Ma, Luo, Wang. "Concatenated and connected random forests with multi scale patch driven active contour model for automated brain tumor segmentation of MR images." *IEEE Transactions on Medical Imaging* 37 no. 8 (2018): 1943–1954.

[10] Singh, Bala. "A DCT-based local and non-local fuzzy C-means algorithm for segmentation of brain magnetic resonance images." *Applied Soft Computing* 68 (2018): 447–457.

[11] Jothi and Hannah Inbarani. "Hybrid tolerance rough set–firefly for MRI brain based supervised feature selection tumor image classification." *Applied Soft Computing* 46 (2016):639–651.

[12] Tong, Zhao, Zhang. "MRI brain tumor segmentation based on texture features and kernel sparse coding." *Biomedical Signal Processing and Control* 47 (2019):387–392.

[13] Lim, Mandava. "A multi-phase semi-automatic approach for multi sequence brain tumor image segmentation." *Expert Systems with Applications* 112 (2018):288–300.

[14] Angulakshmi, Priya. "Brain tumour segmentation from MRI using superpixels based spectral clustering." *Journal of King Saud University-Computer and Information Sciences* (2018). doi:10.1016/j.jksuci.2018.01.009.

[15] Soltaninejad, Yang, Lambrou et al. "Supervised learning based multimodal MRI brain tumour segmentation using texture features from supervoxels." *Computer Methods and Programs in Biomedicine* 157 (2018): 69–84.

[16] Kamathe and Joshi. "A novel method based on independent component analysis for brain MR image tissue classification into CSF, WM and GM for atrophy detection in Alzheimer's disease." *Biomedical Signal Processing and Control* 40 (2018): 41–48.

[17] Pham, Siarry, Oulhadj. "Integrating fuzzy entropy clustering with an improved PSO for MRI brain image segmentation." *Applied Soft Computing* 65 (2018): 230–242.

[18] Li, Jia, Qin. "Brain tumor segmentation from multimodal magnetic resonance images via sparse representation." *Artificial Intelligence in Medicine* 73 (2016): 1–13.

[19] Lahmiri. "Glioma detection based on multi-fractal features of segmented brain MRI by particle swarm optimization techniques." *Biomedical Signal Processing and Control* 31 (2017):148–155.

[20] Alagarsamy, Kamatchi, Govindaraj. "A fully automated hybrid methodology using Cuckoo-based fuzzy clustering technique for magnetic resonance brain image segmentation." *International Journal of Imaging Systems and Technology* 27 no. 4 (2017): 317–332.

[21] Farhi, Yusuf, Raza. "Adaptive stochastic segmentation via energy-convergence for brain tumor in MR images." *Journal of Visual Communication and Image* 46(2017): 303–311.

[22] Ilunga-Mbuyamba, Cruz-Duarte, Avina-Cervantes et al. "Active contours driven by Cuckoo search strategy for brain tumor images segmentation." *Expert Systems with Applications* 56 (2016): 59–68.

[23] Andac, Godze, Kutlay et al. "Tumor-cut: Segmentation of brain tumors on contrast enhanced MR images for radio surgery applications." *IEEE Transactions on Medical Imaging* 31 no. 3 (2012):790–804.

[24] Adhikari, Sing, Basu et al. "Conditional spatial fuzzy C-means clustering algorithm for segmentation of MRI images." *Applied Soft Computing* 34 (2015): 758–769.

[25] Leemput, Colchester, Maes et al. "Automated model based tissue classification of MR images of the brain." *IEEE Transactions on Medical Imaging* 18 no. 10 (1999):897–908.

[26] Qiu, Xiao, Yu et al. "A modified interval type-2 fuzzy C-means algorithm with application in MR image segmentation." *Pattern Recognition Letters* 34 no. 12 (2013):1329–1338.

[27] Moeskops, Manon, Sabina et al. "Automatic segmentation of MR brain images of preterm infants using supervised classification." *Neuro Image* 118 (2015): 628–641.

[28] Markopoulos, Gousias, Ledig. "Automatic whole brain MRI segmentation of the developing neonatal brain." *IEEE Transactions on Medical Imaging* 33 no. 9 (2014): 1818–1831.

[29] Nabizadeh, Kubat. "Brain tumors detection and segmentation in MR images: Gabor wavelet vs. statistical features." *Computers & Electrical Engineering* 45 (2015): 286–301.

[30] Sikka, Sinha, Singh et al. "A fully automated algorithm under modified FCM framework for improved brain MR image segmentation." *Magnetic Resonance Imaging* 27 no. 7 (2009): 994–1004.

[31] Mahmood, Chodorowski, Persson. "Automated MRI brain tissue segmentation based on mean shift and fuzzy c-means using a priori tissue probability maps." *IRBM* 36 no. 3 (2015):185–196.

[32] Govindaraj, Murugan. "A complete automated algorithm for segmentation of tissues and identification of tumor region in T1, T2, and FLAIR brain images using optimization and clustering techniques." *International Journal of Imaging Systems and Technology* 24 no. 4 (2014): 313–325.

18

A Risk Assessment Model for Alzheimer's Disease Using Fuzzy Cognitive Map

S. Meenakshi Ammal and Dr. L. S. Jayashree

Department of Computer Science and Engineering, PSG College of Technology, Coimbatore, Tamil Nadu, India

18.1 Introduction

As per a survey by the WHO, the aging population will reach about 2 billion in 2050 (World Health Organization 2018). Aging-related diseases such as AD have grown rapidly in recent decades and globally 47 million elders are affected by AD, which is projected to reach 75 million in 2030 and 131.5 million in 2050 (Alzheimer's Association 2018a). The report says that early diagnosis of Alzheimer's disease could save $7 trillion to $7.9 trillion in health care costs and 18.4 billion hours of caregivers (Alzheimer's Association 2018b). There are many diagnostic tests used to evaluate the cognitive decline in medical domain. Mini-Mental State Exam (MMSE) is used to check cognitive skills through a series of questions. Mini-Cog is used to evaluate memory power. Mood assessment is used to find out mood disorder. Brain imaging like magnetic resonance imaging (MRI) scan and computed tomography (CT) are used to assess the brain functioning (Alzheimer's Association 2018c).

AD is a progressive and an irreversible disease, which leads to death at the severe stage. Experts suggest that predicting AD is very significant than diagnosing AD. Most of the risk factors for AD could be modified such as physical activity and diet pattern that in turn reduce risk level of AD. To prevent the prevalence of dementia and delay its onset, many prediction techniques have been developed. The cardio-vascular risk factors, aging, and incidence of dementia (CAIDE) is designed for predicting risk of dementia for middle-aged people 20 years early based on vascular risk profile (Sindi et al. 2015). The Australian National University has developed an evidence-based risk prediction tool, Australian National University AD Risk Index (ANU-ADRI), to predict the risk of AD in older patients (Anstey et al. 2014). Brief dementia screening indicator (BDSI) is specifically designed for clinical settings to assist medical expert to identify the increased risk of dementia in older patients (Barnes et al. 2014).

This chapter proposes a fuzzy cognitive map (FCM)-based AD risk assessment model by considering strong risk factors. The rest of this chapter is organized as follows: Section 18.2 describes the related work on FCM-based systems. Section 18.3 describes the FCM-based AD risk assessment model. Section 18.4 illustrates the results obtained and Section 18.5 concludes the paper.

18.2 Related Work

FCM is a powerful soft computing technique to design a human reasoning and inference ability system. FCM-based decision support system has been designed in various domain such as agriculture, structural health monitoring, engineering, medical, etc. Specifically in medical domain, the FCM-based decision support systems have been designed for various purposes (Stylios et al. 2008). A FCM model for radiation therapy (Papageorgiou et al. 2002) is designed for estimating radiation dose in the desired level, which maximizes the radiation dose to the tumor and minimizes the radiation dose to the normal tissues. The case based fuzzy cognitive map (CBFCM) incorporates heterogeneous clinical data from various sources of knowledge into a single clinical practice guidelines (CPG) to design a patient's health record (Douali et al. 2013). CPG covers all the medical history of a patient such as signs, symptoms, biological factor, and genetic data, and it is tested for identifying cardiovascular disease (Douali and Jaulent 2013). FCM-based familial breast cancer (FBM) risk management system is constructed based on demographic risk factors to categorize the individuals into different level of risk (Papageorgiou et al. 2015). A web-based decision-making system is modeled to support dental implantation based on patient's anatomical data (Lee et al. 2012). A diagnosis system in first level and UTI therapy in second level are designed to support clinicians to take correct decision in treatment plan (Papageorgiou et al. 2009). A hierarchical FCM is designed to diagnose speech disorder such as dysarthria and apraxia using speech and language pathologists (Georgopoulos et al. 2005).

The main objective of this paper is to design a decision support system for AD. The proposed system predicts the risk level of AD into low level, moderate level, and high level. The FCM processes the qualitative data efficiently in fuzzy terms better than other machine learning techniques and achieves better diagnostic accuracy.

18.3 AD Risk Assessment Model Using FCM

18.3.1 Overview of FCM

A FCM model (Papageorgiou et al. 2012) is a graphical representation in which nodes denote the concept values and edges represent the influence between the concepts. FCM uses fuzzy rules and the quantity A_i of the input concept takes fuzzy values in the range of [0, 1]. The strength between the concept C_i and C_j is denoted by W_{ij} which ranges from −1 to +1. The concept value of C_i is updated using Eq. (18.1),

$$A_i^{(k+1)} = f\left(A_i^{(k)} + \sum_{\substack{j \neq i \\ j=1}}^{N} A_j^{(k)} \cdot W_{ji}^{(k)}\right) \qquad (18.1)$$

where $A_i^{(k+1)}$ is the updated value of concept C_i at the step $k+1$, $A_j^{(k)}$ and $A_i^{(k)}$ are the values of C_j and C_i at step k. $W_{ji}^{(k)}$ is the strength of C_j and C_i. The sigmoid activation function $f(x)$ is used as given in Eq. (18.2).

$$f(x) = \frac{1}{1 + e^{-\lambda x}} \tag{18.2}$$

where $\lambda = 1$ and $f(x)$ takes the values in the range of $[0, 1]$. Initially, the weight value is derived using linguistic terms such as low, medium, and high. The linguistic terms are converted into single numerical value in the range of $[-1, 1]$ by the defuzzification method. The proposed AD risk assessment model uses nonlinear Hebbian learning (NHL) rule (Papageorgiou 2012) to update the strength between the concepts as given in Eq. (18.3).

$$W_{ji}^{(k+1)} = W_{ji}^{(k)} + \eta_k A_j^{(k)} \left(A_i^{(k)} - A_j^{(k)} W_{ji}^{(k)} \right) \tag{18.3}$$

where η_k is learning rate, $W_{ji}^{(k+1)}$ and $W_{ji}^{(k)}$ are the weight values at step $k + 1$ and k. The termination rules followed by NHL are given in Eqs. (18.4) and (18.5).

$$F1 = \sqrt{DC_j^{(k)} - T_j^2} \tag{18.4}$$

$$F2 = \left| DC_j^{k+1} - DC_j^{(k)} \right| < e \tag{18.5}$$

where $DC_j^{(k)}$ is the actual outcome, T_j is the target outcome, and $e = 0.001$.

18.3.2 FCM Model for AD Risk Assessment

The proposed risk assessment model uses various strong risk factors that cause the onset of AD in future. The 10 AD risk factors (Alzheimer's Association 2018a; Alzheimer Scotland 2011; Alzheimer Society Canada 2018) are identified using medical literature and are used as input concepts for the FCM model. The risk factors (Alzheimer's Association 2018b; Alzheimer Scotland 2011; Alzheimer Society Canada 2018) are described in Table 18.1.

TABLE 18.1

Risk Factors for AD

Concepts	Risk Factors for AD	Description
C1	Age	Old age (above 60 years) is the strongest risk factor for developing AD.
C2	Family history	Family history is the main risk factor on the prevalence of dementia.
C3	Risk genes	The number of copies of ApoE gene inherited increases the risk of AD.
C4	Blood pressure	High blood pressure has a strong association with cognitive decline.
C5	Type-2 diabetes	Diabetes, especially, type-2 diabetes is a strong risk factor of AD.
C6	Physical activity	Physical activity increases the brain functionality and reduces the risk of AD.
C7	Alcohol consumption	Taking excess level of alcohol increases the chance of AD.
C8	Smoking	Smoking is the strong risk factor of AD.
C9	Obesity (BMI)	Excessive body weight leads to various health issues and increase the chance of AD.
C10	Unhealthy diet	Taking unhealthy food increases the risk of AD.

The range of values for input concept is derived from clinical experts, online medical resources, and medical literature. Generally, the risk factors for AD are not quantifiable and imprecise. FCM enables to model those features using fuzzy terms. The fuzzy values for each input concept are given in Table 18.2.

The influences of each input concept to yield final output concept are derived by fuzzy if-then rules. The 11th concept is the final output concept termed as AD risk grade (ADRG). The three experts gave their opinion about the weight between the input and output concept as follows:

First Expert opinion
    If the value (age in years) of C1 is high, then the AD risk grade C11 is high.
    Inference: The causal relationship between C1 and C11 is high.

Second Expert Opinion
    If the value of C1 is high, then the AD risk grade C11 is medium.
    Inference: The causal relationship between C1 and C11 is medium.

Third Expert Opinion
    If the value of C1 is medium, then AD risk grade C11 is low.
    Inference: The causal relationship between C1 and C11 is low.

The causal relationship between the concepts is derived from three clinical experts in linguistic terms. Experts' opinions are aggregated using SUM method to derive a single linguistic variable. Then the single linguistic variable is converted into a numerical weight value using defuzzification of center of gravity (CoG). Following this method,

TABLE 18.2

Fuzzy Values for AD Risk Factor

Concepts	Range of Fuzzy Values		
	Low	Moderate	High
C1:Age The age of the patient in years	60–75	75–85	85–95
C2: Family history The number of family members having AD	0	1	≥ 2
C3: Risk genes The number of copies of APOE type genes inherited	0	1	2
C4: Blood pressure It is measured in mm Hg.	>90/60 & <120/80	>120/80 & <140/90	>140/90
C5: Type 2 diabetes It is measured in mg/dL.	< 100	100–125	≥ 126
C6: Physical activity Physical activity in hours per day	1	0.30	0
C7: Alcohol consumption The number of times alcohol consumed per day	0	1	≥ 2
C8: Smoking The number of smoking per day	0	1	≥ 2
C9: Obesity It is measured in kg/m^2.	18.5–24.9	25.0–29.9	30.0–39.9
C10: Unhealthy diet It is measured by taking unhealthy food per week.	1	2	≥ 3

the relationship between the input and output concept is calculated and the weight matrix W is generated as given in Figure 18.1.

The CoG method creates a numeric weight value in the range of [1,−1]. The FCM model for the proposed AD risk assessment is illustrated in Figure 18.2.

$$
W_{initial} = \begin{vmatrix}
0 & 0 & 0 & 0 & 0 & 0 & 0 & 0 & 0 & 0 & 0.82 \\
0 & 0 & 0 & 0 & 0 & 0 & 0 & 0 & 0 & 0 & 0.57 \\
0 & 0 & 0 & 0 & 0 & 0 & 0 & 0 & 0 & 0 & 0.68 \\
0 & 0 & 0 & 0 & 0 & 0 & 0 & 0 & 0 & 0 & 0.56 \\
0 & 0 & 0 & 0.57 & 0 & 0 & 0 & 0 & 0 & 0 & 0.78 \\
0 & 0 & 0 & 0 & 0 & 0 & 0 & 0 & -0.65 & 0 & 0.52 \\
0 & 0 & 0 & 0 & 0.75 & 0 & 0 & 0 & 0 & 0 & 0.42 \\
0 & 0 & 0 & 0 & 0 & 0 & 0 & 0 & 0 & 0 & 0.45 \\
0 & 0 & 0 & 0 & 0 & 0 & 0 & 0 & 0 & 0 & 0.52 \\
0 & 0 & 0 & 0 & 0 & 0 & 0 & 0 & 0.72 & 0 & 0.39 \\
0 & 0 & 0 & 0 & 0 & 0 & 0 & 0 & 0 & 0 & 0
\end{vmatrix}
$$

FIGURE 18.1
Initial Weight Matrix Created by Experts.

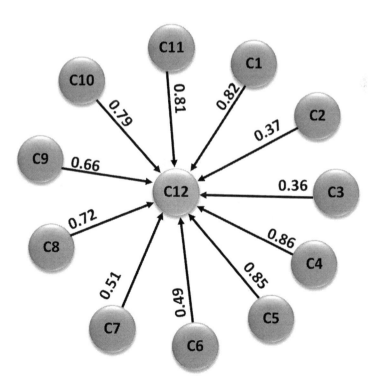

FIGURE 18.2
FCM for AD Risk Factor.

18.3.3 AD Risk Assessment Using NHL-FCM

The proposed AD risk assessment model is trained using NHL algorithm and predicts the risk level of AD for the given input risk factors. Initially, the weights of the model are assigned using expert's opinion. The normalized form of input risk factors is obtained using Eq. (18.6). The NHL updates the weight using Eqs. (18.1) through (18.3) until the input and output concept reaching the desired condition using Eqs. (18.4) and (18.5).

$$X_{norm} = \frac{X - X_{min}}{X_{max} - X_{min}} \tag{18.6}$$

The learning rate initially assigned was 0.01 and minimized exponentially at each iteration step k using Eq. (18.7).

$$\eta^{(k)} = \frac{\eta^{(k-1)}}{(2k+1)} \tag{18.7}$$

The convergence phase is attained according to Eqs. (18.4) and (18.5) and the output concept value at the convergence stage grants the risk level of AD. Due to experts' opinion, the output concept C11 is defined in the range of 0 to 0.35 for low risk, 0.35 to 0.75 for moderate risk, and 0.75 to 1 for high risk. The proposed AD risk assessment using NHL algorithm is tested using 30 sample records. The results obtained show that the NHL algorithm efficiently categorizes risk level of AD and assists the individual to predict the risk of AD using their risk profile. The AD risk assessment generates the risk grade, which is concurrent with the expert's opinion and medical literature.

18.4 Results and Discussion

The results obtained using NHL-based AD risk assessment are explained for three different risk levels of low risk, moderate risk, and high risk as follows:

18.4.1 Scenario A: Low Risk for AD

The input risk factors are chosen from Table 18.2 as follows:

C1 is 67 years, C2 is one relative having AD, C3 is one copy inherited, C4 has 1.43 mm Hg, C5 has 90 mg/dL, C6 is 0.15 hours per day, C7 is one drink per day, C8 is one smoke per day, C9 is 17 kg/m^2, C10 is one time unhealthy food per week.

These input concept values are normalized using Eq. (18.6) to obtain the value in the range of 0–1 as follows:

$$A^1_{initial} = \begin{array}{cccccc} [0.097 & 0.250 & 0.500 & 0.142 & 0.125 & 0.15 \\ 0.212 & 0.251 & 0.016 & 0.213 & 0.000] \end{array}$$

The normalized values are processed using NHL algorithm using Eqs. (18.1)–(18.3). The weight updating is continued until the conditions in Eqs. (18.4) and (18.5) are satisfied. The learning rate was assigned as $\eta = 0.01$. After 10 iterations the values are converged in the specified region as follows:

$$A^1_{final} = \begin{bmatrix} 0.231 & 0.282 & 0.262 & 0.189 & 0.264 & 0.283 \\ 0.282 & 0.262 & 0.194 & 0.268 & 0.272 \end{bmatrix}$$

The final output concept value is 0.272, which is converged after 10 iterations in a stable state. The value in the range of 0–0.35 indicates *low risk* level of AD.

18.4.2 Scenario B: Moderate Risk for AD

The input risk factors are chosen which generate moderate risk as follows:
C1 is 73 years, C2 is two relatives having AD, C3 is one copy inherited, C4 has 1.56 mm Hg, C5 has 50 mg/dL, C6 is 0.30 hours per day, C7 is two drinks per day, C8 is two smokes per day, C9 is 27 kg/m², C10 is three times unhealthy food per week.

The input concept values are normalized using Eq. (18.6) and the values are obtained as follows:

$$A^2_{initial} = \begin{bmatrix} 0.290 & 0.50 & 0.50 & 0.765 & 0.676 & 0.300 \\ 0.400 & 0.500 & 0.303 & 0.600 & 0.000 \end{bmatrix}$$

After 12 iterations, the output concept value is derived as follows:

$$A^2_{final} = \begin{bmatrix} 0.641 & 0.565 & 0.662 & 0.678 & 0.676 & 0.654 \\ 0.631 & 0.578 & 0.683 & 0.679 & 0.678 \end{bmatrix}$$

The output concept value is in the range of 0.35–0.75, which indicates *moderate risk* of AD.

18.4.3 Scenario C: High Risk for AD

The input factors which yield high risk of AD are chosen from Table 18.2 and are given as follows:
C1 is 95 years, C2 is three relatives having AD, C3 is two copies inherited, C4 has 1.57 mm Hg, C5 has 40 mg/dL, C6 is 0 hours per day, C7 is four drinks per day, C8 is four smokes per day, C9 is 45 kg/m², C10 is four times unhealthy food per week.

The input risk factors are normalized and the values are derived as follows:

$$A^3_{initial} = \begin{bmatrix} 0.976 & 0.750 & 0.871 & 0.824 & 0.956 & 0.01 \\ 0.800 & 0.876 & 0.839 & 0.848 & 0.000 \end{bmatrix}$$

The input values are converged after nine iterations and the output concept is derived in the range of 0.75–1 as follows:

$$A^3_{final} = \begin{bmatrix} 0.785 & 0.854 & 0.863 & 0.782 & 0.876 & 0.848 \\ 0.753 & 0.746 & 0.866 & 0.823 & 0.972 \end{bmatrix}$$

Figure 18.3 illustrates the convergence of 11 concept values in each step.

The sample records tested using NHL algorithms have brought the reliable outcome as shown in Table 18.3. The results obtained using NHL algorithm show that the value of output concept has converged in the specified range, which is concurrent with the expert's opinion.

The 96% diagnostic accuracy is attained using FCM-based nonlinear Hebbian learning algorithm. Out of 30 sample records, 29 samples are classified correctly. Nine samples for low risk have converged in the region of 0–0.35 and achieve 100% of classification

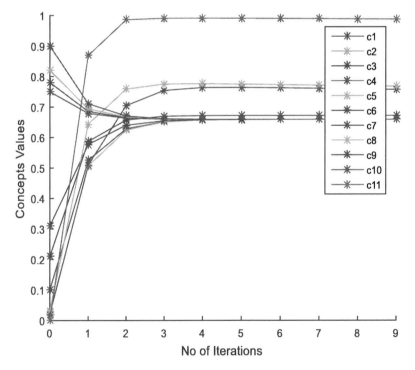

FIGURE 18.3
Convergence of Concept Values Using NHL.

TABLE 18.3

AD Risk Assessment Using FCM-NHL

S. No.	C1	C2	C3	C4	C5	C6	C7	C8	C9	C10	AD Risk Grade (Proposed NHL-FCM)
1.	70	0	0	1.31	92	1	0	0	18.5	1	Low
2.	72	1	0	1.33	93	1	0	0	19.2	1	Low
3.	71	0	0	1.32	94	1	0	0	20.5	1	Low
4.	74	1	0	1.21	95	1	0	0	18.2	1	Low
5.	65	0	0	1.52	96	1	0	0	18.5	1	Low
6.	60	1	0	1.33	97	1	0	0	19.8	1	Low
7.	62	2	0	1.51	98	1	0	0	23.4	1	Low
8.	70	3	0	1.52	91	1	0	0	24.8	1	Low
9.	73	2	0	1.42	92	1	0	0	23.6	1	Low
10.	64	1	0	1.31	94	1	0	0	24.6	1	Moderate
11.	78	0	1	1.32	101	0.30	1	1	25	2	Moderate
12.	76	3	1	1.33	104	0.30	1	1	25.7	2	Moderate
13.	77	2	1	1.34	106	0.32	1	1	26.6	2	Low
14.	60	1	1	1.35	110	0.32	1	1	27.6	2	Moderate
15.	81	0	1	1.36	120	0.33	1	1	27.6	2	Moderate
16.	82	1	1	1.41	125	0.30	1	1	28	2	Moderate
17.	83	3	1	1.42	123	0.30	1	1	27.3	2	Moderate
18.	84	1	1	1.43	124	0.30	1	1	26.1	2	Moderate
19.	82	2	1	1.45	106	0.30	1	1	26.6	2	Moderate
20.	80	3	1	1.46	108	0.30	1	1	29	2	Moderate
21.	85	0	2	1.52	127	0	2	2	30.4	3	High
22.	90	0	2	1.54	128	0	2	2	32.5	3	High
23.	94	1	2	1.51	129	0	2	2	31.2	3	High
24.	91	1	2	1.53	130	0	2	2	32.8	4	High
25.	95	2	2	1.52	134	0	2	2	33.1	4	High
26.	91	3	2	1.51	128	0	3	2	34.5	4	High
27.	87	3	2	1.42	126	0	3	3	36.7	4	High
28.	88	1	2	1.42	125	0	3	3	33.4	3	High
29.	86	2	2	1.31	127	0	3	3	35.6	3	High
30.	85	1	2	1.32	130	0	3	3	39.5	3	High

accuracy. Out of 11 samples, 10 samples for moderate risk have converged in the region of 0.35–0.75 and achieve 90% of classification accuracy. Ten samples for high risk have converged in the region of 0.75–1 and attains 100% of classification accuracy. The proposed model is compared with other machine learning methods and generates better diagnostic accuracy of 96% than other methods, which are illustrated in Table 18.4. The proposed AD risk assessment using NHL-based algorithm can be trained even with a small dataset to predict the risk of AD with high accuracy. However, the proposed model should be validated with the real patient record to evaluate system reliability.

TABLE 18.4

Classification Accuracy of Machine Learning Techniques

Classification Techniques	AD Severity	Confusion Matrix			Classification Accuracy (%)
		Low	**Moderate**	**High**	
Proposed model	Low	9	0	0	96
	Moderate	1	10	0	
	High	0	0	10	
MLP	Low	8	1	0	93
	Moderate	0	11	0	
	High	0	1	9	
SVM	Low	7	2	0	86
	Moderate	2	9	0	
	High	0	0	10	
Naïve Bayes	Low	6	3	0	80
	Moderate	2	9	0	
	High	0	1	9	

18.5 Conclusion

AD, the most common form of dementia found in elders, is a progressive and an irreversible disease, which leads to death in severe stage. As there is no proper medication to stop the progression of AD, the early diagnosis could only prevent or delay the onset of AD. The system modeling using risk factors justifies the use of FCM, which represents the features in fuzzy terms. The proposed system categorizes the AD risk into low, moderate, and high and also predicts the risk of AD for elders. The flexibility of the proposed system facilitates to include the new risk factors and updates the system functionality. The proposed system can be further improved to attain high reliability and usability for assisting clinical expert in their treatment process.

References

Alzheimer Scotland. "Risk factors in dementia." 2011. www.alzscot.org/information_and_resources/information_sheet/1786_risk_factors_in_dementia.

Alzheimer Society Canada. "Risk factors." 2018. http://alzheimer.ca/en/Home/About-dementia/Alzheimer-s-disease/Risk-factors.

Alzheimer's Association. "Alzheimer's and Dementia." (2018a). www.alz.org/alzheimer_s_dementia

Alzheimer's Association. "Medical Tests." 2018b. www.alz.org/alzheimers-dementia/diagnosis/medical_tests.

Anstey KJ, Cherbuin N, Herath PM, et al. "A self-report risk index to predict occurrence of dementia in three independent cohorts of older adults: The ANU-ADRI." PLoS One 9(1) (2014): e86141. doi:10.1371/journal.pone.0086141.

Barnes DE, Beiser AS, Lee A, et al. "Development and validation of a brief dementia screening indicator for primary care." Alzheimer's and Dementia 10(6) (2014): 656–665.e1. doi:10.1016/j.jalz.2013.11.006.

Douali N, Jaulent MC. "Integrate personalized medicine into clinical practice to improve patient safety." IRBM 34(1) (2013): 53–55. doi:10.1016/j.irbm.2013.01.001.

Georgopoulos VC, Malandraki GA. "A fuzzy cognitive map hierarchical model for differential diagnosis of dysarthria and apraxia of speech." 2005 IEEE Engineering in Medicine and Biology 27th Annual Conference 3 (2005): 2409–2412. doi:10.1109/IEMBS.2005.1616954.

Papageorgiou EI, Chris Papadimitriou KS. "Management of uncomplicated urinary tract infections using fuzzy cognitive maps." 2009 9th International Conference on Information Technology and Applications in Biomedicine (2009): 1–4. doi:10.1109/ITAB.2009.5394374.

Lee S, Yang J, Han J. "Development of a decision making system for selection of dental implant abutments based on the fuzzy cognitive map." Expert Systems with Applications 39(14) (2012): 11564–11575. doi:10.1016/j.eswa.2012.04.032.

Papageorgiou E, Stylios CD, Groumpos PP. "Decision making in external beam radiation therapy based on fuzzy cognitive maps." Proceedings First International IEEE Symposium Intelligent Systems (2002):320–325. doi:10.1109/IS.2002.1044275.

Papageorgiou EI. "Learning algorithms for fuzzy cognitive maps – A review study." IEEE Transactions on Systems, Man, and Cybernetics, Part C: Applications and Reviews 42(2) (2012): 150–163. doi:10.1109/TSMCC.2011.2138694.

Papageorgiou EI, Subramanian J, Karmegam A, et al. "A risk management model for familial breast cancer: A new application using fuzzy cognitive map method." Computer Methods and Programs in Biomedicine 122(2) (2015): 123–135. doi:10.1016/j.cmpb.2015.07.003.

Sindi S, Calov E, Fokkens J, et al. "The CAIDE dementia risk score app: The development of an evidence-based mobile application to predict the risk of dementia." Alzheimer's & Dementia (Amst) 1(3) (2015): 328–333. doi:10.1016/j.dadm.2015.06.005.

Stylios CD, Georgopoulos VC, Malandraki GA, Chouliara S. "Fuzzy cognitive map architectures for medical decision support systems." Applied Soft Computing 8(3) (2008): 1243–1251. doi:10.1016/j.asoc.2007.02.022.

World Health Organization. "Ageing and health." (2018). www.who.int/news-room/fact-sheets/detail/ageing-and-health.

19

Comparative Analysis of Texture Patterns for the Detection of Breast Cancer Using Mammogram Images

J. Shiny Christobel

Department of ECE, Sri Ramakrishna Institute of Technology, Coimbatore, Tamil Nadu, India

J. Joshan Athanesious

Madras Institute of Technology, Chennai, Tamil Nadu, India

19.1 Introduction

After HIV, breast cancer is the second leading cause of death among women all over the world. About 10% of women suffer from this disease during their lifetime [1]. Cancer is the spontaneous growth of cells that originate in the blood tissue. Tumors may be noncancerous (benign) or cancerous (malignant) tumors which destroys a particular part of the body completely. Classification of benign and malignant tumors will be accurate only if detected early [2]. In the year 2012, about 70,218 women died in India due to breast cancer, more than any other country in the world (second is China with 47,984 deaths and third the United States with 43,909 deaths). About 25% to 31% women die due to breast cancer in India every year. Mammographic screening is widely used in early detection of breast cancer. Through mammogram analysis, radiologists have a detection rate of 76% to 96% which is considerably higher than clinical examination. Many electronic databases are available for mammographic images [3]. In this chapter, we have taken images from Mammographic Image Analysis Society (MIAS) database.

Textures are one of the important characteristics for identifying objects and region of interest of various kinds of images [4]. Texture analysis is important for computer image analysis for classification and detection segmentation of an image based on intensity and color [5, 6]. The texture of a region describes the pattern of spatial variation of gray tones in a neighborhood where the neighborhood is small compared to the region [7].

The evaluation of data taken from patient and decisions of experts are the most important factors in diagnosis. Early detection and accurate diagnosis of the disease is a key factor. Ultrasound does not use radiation. The use of ultrasound can increase overall cancer detection by 17% and reduce the number of unnecessary biopsies by 40%. Breast ultrasound (BUS) imaging is superior to mammography. It is used to get a picture of the inside of the breasts. It is also a good tool which provides more information of the cyst. But there are also certain drawbacks in it. Ultrasound is operator dependent and so if the technician is not well trained, we will not be able to get the actual details of the cyst. To avoid all such problems, we go for mammogram detection. The principle feature on a mammogram is the breast border, otherwise known as skin–air interface. Digital mammogram is used to partition the breast contour into breast and nonbreast region.

Local binary pattern (LBP) is a unique approach for texture classification. LBP was first proposed by Ojala et al. [8]. This approach is simple and efficient and provides detailed image and attractive classifications of results. Many applications such as face recognition [9], dynamic texture recognition [10], and shape localization [11] use this approach. LGP is the other method for texture classification. Here, the local pattern is computed based on the local gradient flow from one side to another side through a 3×3 center pixel [12]. The use of classifiers in medical diagnosis is increasing rapidly. Here, we use support vector machine (SVM) classifiers. SVM classifiers are used to classify benign and malignant tumors. SVM classification is considered as effective statistical learning methods for classification. They rely on support vector for classification [13]. In this chapter, we have modified the existing LBP and LGP. This will reduce the size of pixel and increases the accuracy.

19.2 Methodology

Figure 19.1 shows the block diagram for the detection of breast cancer.

Figure 19.1 shows the mammogram image which is the X-ray image, and preprocessing is used to enhance the image and reduce noise without destroying the important

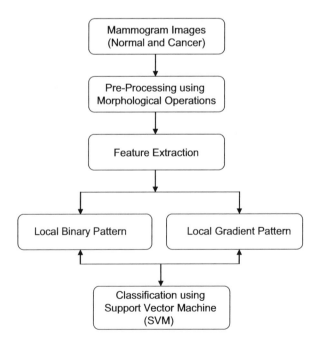

FIGURE 19.1
Block Diagram.

features of the image for diagnosis [14]. The texture of a region describes the pattern of spatial variation of gray tones in a neighborhood where the neighborhood is small compared to the region. Compared to ANN classifier, the process of SVM is 700 times faster and accurate [15]. SVM contains two processes: (1) training and (2) testing. In training, the mean value is computed for the images and the values are trained for benign and malignant tumors [16]. With the help of these trained values, the input images are tested for tumors [17].

19.2.1 Local Binary Pattern

Local binary pattern is a simple yet efficient operator to achieve impressive results in classification of breast cancer. LBP is used to transform mammogram image into texture image. The center pixel is stipulated to 3 × 3 matrix which represents the image of the gray scale and it converts the binary values to eight bit decimal code [18]. Figure 19.1(a) shows the input image in which the center value (say 35) in a matrix is taken as a center pixel and it compares with the other pixels. If the center pixel is greater than other pixels, then the value will be 1 and if it is less, then the value will be 0.

Now, all the binary values are gathered in a sequential order and an equivalent decimal code is generated.

Similarly, all the pixels in the matrix are considered as a center pixel and a new matrix is created with the generated decimal codes. Mean and variance are calculated for the newly generated matrix [19]. This value is sent into the SVM classifier where the mean and variance of the classifier are compared with value of the newly created matrix. The classifier classifies the tumors as benign or malignant according to the compared mean and the variance value. The histograms are computed and used as a texture descriptor [20].

19.2.2 Local Gradient Pattern

The local gradient pattern computes the average value of the eight pixels leaving the center pixel. This average value is taken as the threshold value and compared with the neighboring pixels. The pixel value is taken as 1 if the gradient value of the neighboring

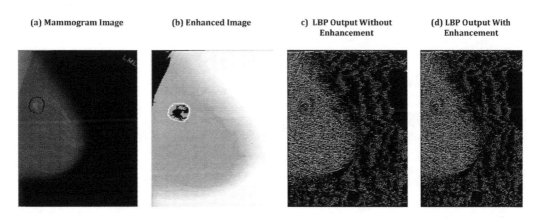

(a) Mammogram Image   (b) Enhanced Image   c) LBP Output Without Enhancement   (d) LBP Output With Enhancement

FIGURE 19.2
LBP Output.

pixel is greater than the threshold value and as 0 if it lesser than the threshold value [21]. The binary values are generated. These values are converted to eight bit decimal code and stored in a new matrix. This procedure is repeated for the other pixels and the new matrix is generated. This matrix is sent to the SVM classifier for classification [22]. For example, Figure 19.2(a) shows the input image in which the average value is computed for all the eight pixels except the center pixel.

The average is taken as the threshold value and compared with the neighboring pixels.

Now, all the binary values are gathered in a sequential order and an equivalent eight bit decimal code is generated.

19.3 Results and Discussion

19.3.1 Local Binary Pattern Output

The enhanced images of LBP and LGP are obtained by morphological operations like dilation, erosion, opening, closing, top-hat transform, and bottom-hat transform. The contrast image is the resultant image obtained by the morphological operations. The complement of the contrast image is the output of the enhanced image and shown in Figure 19.2(b). The output of LBP with enhancement and without enhancement is shown in Figure 19.2(c) and (d).

19.3.2 Local Gradient Pattern Output

The outputs of LGP with enhancement and without enhancement are shown in Figure 19.3(c) and (d).

In this chapter, the SVM classifier is trained to classify and detect the cancer images with more accuracy. The modified LBP and LGP reduce the computational time and increases accuracy. Table 19.1 shows the number of mammogram images that are trained and tested by the SVM classifier and also the number of detected images for modified LBP and LGP with and without enhancement.

Calculate accuracy, sensitivity, specificity using (9.1), (9.2), (9.3), and (9.4) [16].

$$\text{Accuracy } (T) = \frac{\sum_{i=1}^{|T|} \text{assess}(t_i)}{|T|} t_i \in T \tag{9.1}$$

$$\text{Assess } (t) = \begin{cases} 1, & \text{if classify}(t) \equiv \text{correct classification} \\ 0, & \text{otherwise} \end{cases} \tag{9.2}$$

where T is the total number of data to be classified.

$$\text{Sensitivity} = \frac{TP}{TP + FN}(\%) \tag{9.3}$$

$$\text{Specificity} = \frac{TN}{FP + TN}(\%) \tag{9.4}$$

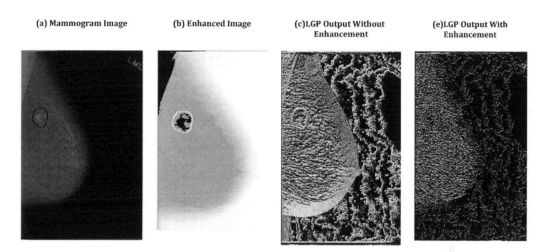

FIGURE 19.3
LGP Output.

TABLE 19.1

LBP and LGP Results

			LBP		LGP	
SVM Classifier			**Detected Images**			
Types of Images	**Trained Images**	**Tested Images**	**With Enhancement**	**Without Enhancement**	**With Enhancement**	**Without Enhancement**
Normal	300	100	90	89	93	94
Abnormal	300	100	92	86	91	93

where TP is true positive, TN true negative, FP is false positive, and FN is false negative.

True positive (TP): An input is detected as a cancer by the SVM classifier.

True negative (TN): An input is detected as a normal and labeled as a healthy person by the SVM classifier.

False positive (FP): An input is detected as a cancer and labeled as a healthy person by the SVM classifier.

False negative (FN): An input is detected as a normal by the SVM classifier but with cancer.

TABLE 19.2

Performance Measures

	Accuracy	Sensitivity	Specificity
LBP with enhancement	83.33%	80%	86.66%
LBP without enhancement	55.6%	20%	90%
LGP with enhancement	85%	80%	91%
LGP without enhancement	80%	80%	80%

Compared to LBP and LGP, LGP shows more accuracy in detecting cancer and normal images correctly shown in Table 19.2.

19.4. Conclusion

In this chapter, we compare LBP with LGP. From the outputs mentioned before, we conclude that LGP is more accurate and coherent than LBP. The X-axis represents the patterns and the Y-axis represents the count of the histogram. LGP shows peak patterns than LBP. And it has smaller detection error than LBP because the left and right overlapping region corresponds to false negative detection error and the false positive detection error. Thus, we conclude that LGP with enhancement is simpler, accurate, and coherent compared to LBP.

References

[1] Cheng, H. D., J. Shan, W. Ju, Y. Guo, and L. Zhang, "Automated breast cancer detection and classification using ultrasound images: A survey," *Pattern Recognition*, vol. 43, pp. 299–317, 2010.

[2] Akay, M. F., "Support vector machine combined with feature selection for breast cancer diagnosis," Expert Systems with Applications Pattern Recognition, vol. 36, pp. 3240–3247, 2009.

[3] Cahoon, T. C., M. A. Sutton, and J. C. Bezdek, "Breast cancer detection using image processing technique."

[4] Xizhaoli, S. W., and M. J. Bottem, "A texture and region dependent breast cancer risk assessment from screening mammogram," Pattern Recognition Letters, vol. 36, pp. 117–124, 2014.

[5] Haralick, R. M., K. Shanmugam, and I. Dinstein, "Textural features for image classification," IEEE Transaction on System Man and Cybernetics, vol. 3, no. 6, pp. 610–621, November 1973.

[6] Ojala, T., K. Valkealahti, E. Oja, and M. Pietikainen, "Texture discrimination with multi-dimensional distributions of signed gray level differences," Pattern Recognition, vol. 34, no. 3, pp. 727–739, 2001.

[7] Hawlick, R. M., and S. Member, "Statistical and structural approaches to texture," Proceedings of IEEE, vol. 67, no. 5, May 1979.

[8] Ojala, T., M. Pietikäinen, and T. T. Mäenpää, "Multiresolution gray-scale and rotation invariant texture classification with local binary pattern," IEEE Transactions on Pattern Analysis and Machine Intelligence, vol. 24, no. 7, pp. 971–987, 2002.

[9] Ahonen, T., A. Hadid, and M. Pietikäinen, "Face recognition with local binary patterns: Application to face recognition," IEEE Trans. On Pattern Analysis and Machine Intelligence, vol. 28, no. 12, pp. 2037–2041, 2006.

[10] Zhao, G., and M. Pietikainen, "Dynamic texture recognition using local binary patterns with an application to facial expressions," IEEE Transactions on Pattern Analysis and Machine Intelligence, vol. 27, no. 6, pp. 915–928, 2007.

[11] Huang, X., S. Z. Li, and Y. Wang, "Shape localization based on statistical method using extended local binary pattern," in Proceedings of International Conference on Image and Graphics, 2004, pp. 184–187.

[12] Jun, B., and D. Kim, "Robust face detection using local gradient pattern and evidence accumulation," Pattern Recognition, vol. 45, pp. 3304–3316, 2012.

[13] Wang, D., P. LinShi, and A. Heng, "Automatic detection of breast cancers in mammograms using structured support vector machine," Neurocomputing, vol. 72, pp. 3296–3302, 2009.

[14] Chang, R. F., W. J. Wu, W. K. Moon, and D. R. Chen, "Improvement in breast tumour discrimination by support vector machines and speckle-emphasis texture analysis," Ultrasound in Medicine and Biology, vol. 29, no. 5, pp. 679–686, 2003.

[15] Xiangjun Shi, H. D. C., and L. Hu, "Mass detection and classification in breast ultrasound images using fuzzy SVM," in JCIS-2006 Proceedings, 2006.

[16] Yu, H., and G. Rumbe, "Comparative study of classification techniques on breast cancer FNA biopsy data: a direct path to intelligent tools."

[17] Polat Salih Güne, K., "Breast cancer diagnosis using least square support vector machine," Digital Signal Processing, vol. 17, pp. 694–701, 2007.

[18] Guo, Z., L. Zhang, I. E. E. E. Member, D. Zhang, and I. E. E. E. A. Fellow, "Completed modelling of local binary pattern operator for texture classification," Submitted to IEEE Transactions on Image Processing.

[19] Llado, X., A. Oliver, J. Freixenet, R. Martí, and J. Marti, "A textural approach for mass false positive reduction in mammography," Computerized Medical Imaging and Graphics, vol. 33, pp. 415–422, 2009.

[20] Sabu, A. M., D. N. Ponraj, and Poongodi, "Textural features based breast cancer detection: A survey," Journal of Emerging Trends in Computing and Information Sciences, vol. 3, no. 9, September 2012.

[21] Islam, M. S., "Local gradient pattern – A novel feature representation for facial expression recognition," The International Journal of Artificial Intelligence & Data Mining (JAIDM).

20

Analysis of Various Color Models for Endoscopic Images

Caren Babu

Karunya Institute of Technology and Sciences, Coimbatore, Tamil Nadu, India

Anand Paul

Kyungpook National University, Daegu, South Korea

D. Abraham Chandy

Karunya Institute of Technology and Sciences, Coimbatore, Tamil Nadu, India

20.1 Introduction

The endoscopic examination of gastrointestinal (GI) tract plays a vital role in the diagnosis of various diseases associated with the GI tract. The endoscopic procedures such as upper GI endoscopy, colonoscopy, and enteroscopy can visualize the entire GI tract depending upon the procedure involved. The endoscopic procedure used in this inspection can be either wired or wireless. With the development of an ingestible capsule into the body of the patient, wireless capsule endoscopy (WCE) has proven itself to be a promising procedure to record the internal images of the GI tract. During the past decade, WCE has succeeded in achieving major improvements and thus a valuable alternative to its wired counterpart. Currently, several capsules (Olympus, Pillcam, Endocapsule, OMOM, Microcam) are available in the market [1]. The endoscopic videos are captured in real time which takes 8–20 hours and include more than 50,000 images which are sent wirelessly to the endoscopic data recorder situated outside the body.

The endoscopic videos/images contain information about the conditions of the intestinal mucosa, expressed by color, texture, and pattern. These parameters are extremely important for the diagnosis and therefore, should be preserved with utmost care. The majority of the image sensors of digital cameras available today, including the sensors employed for WCE, are based on the RGB model. However, the constraints on various image processing would demand some application specific modifications which may be a preprocessing operation on such images. Figure 20.1 shows the sample images taken from different classes of the experimental dataset. The endoscopic images compared to the standard images like Baboon are observed to be having several unique properties such as homogeneity in color, the dominance of red components, the absence of sharp edges, etc. The presence of a limited number of colors in the endoscopic images paved a way for color conversion. The high correlation

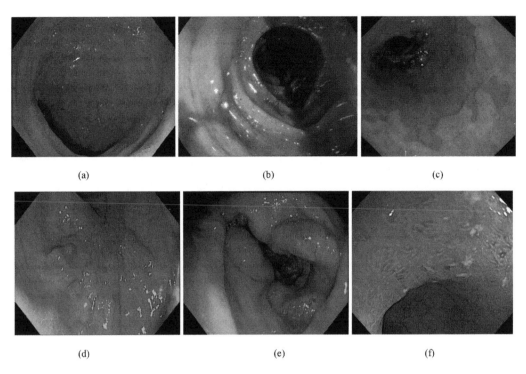

(a)   (b)   (c)

(d)   (e)   (f)

FIGURE 20.1
Sample endoscopic images of various categories in the dataset: (a) Lumen (image no. 18), (b) contents (image no. 32), (c) mucosa (image no. 50), (d) flat lesion (image no. 64), (e) protruding lesions (image no. 87), (f) excavated lesions (image no. 127). (Source: WCE Clinical Endoscopy Atlas [21]).

between the component values is a severe drawback of the RGB model that motivates researchers to focus more on other models [2]. Researches show that if the algorithm can access the image stored directly in luminance (brightness) and color component, the processing can be made faster. In addition, the availability of simple conversion formula from the RGB model to such color space and vice versa without visible losses in the data increased the scope of color conversion.

The color models represent the color information in a three-dimensional (3D) space with each pixel having a 3D color vector. It is possible to represent any color image in terms of its components. There are several color spaces available in the literature that is applicable to endoscopic images which vary from application to application [3]. In recent years with the advancement in WCE, the demands for low power processing inside the capsule and a suitable compression module applicable to endoscopic images arise. In addition, there is a growing trend to record and store videos of endoscopic procedures, mainly for medical documentation and research. YUV color space is used in compressed sensing [4] as inbuilt RGB to YUV converters are present in several capsules [5] and thus reduce the burden on hardware. A recent study has shown that YEF color space has superior performance than YUV and YCgCo leads to its application in ref. [6]. Another color space, YCbCr, is also prominent in compression application. In addition to compression application [7, 8], the YCbCr color space is also used for detecting bleeding frames in refs. [9, 10] as a color enhancement method for endoscopic images.

In ref. [11], YCbCr has been replaced by YCgCo considering the power constraints. In ref. [2], detection of bleeding frames using the HSV model is discussed.

Another prominent application of endoscopic image in literature is abnormality detection. The automatic detection of abnormalities using endoscopic image is a growing field of interest for biomedical researchers. The texture-based image segmentation and pathology detection using the YIQ color model show superior results in ref. [12]. The YIQ is also used for the detection of GI adenomas in video endoscopy by applying a color transformation in the video frame which improves the texture discrimination [13]. In ref. [12], YUV is used along with YIQ for abnormality detection whereas in ref. [14] HSI color space is used by performing segmentation on endoscopic images. There are mainly two advantages for HSI color space. Firstly, HSI is suitable for specifying color in a more suitable way for human perception. Secondly, this model can control the intensity and chromatic components more easily and independently. It should be noted that HSI and HSV color space shows superior results for endoscopic image classification system mainly due to its relevance in perception [15]. Another color model CIE Lab is used for segmentation of clinical endoscopic images into integral parts to record a sequence of only interesting parts [16]. The application of color models for endoscopic images extends to image steganography to maintain secure communication of medical data. In ref. [17], HSV color space is suggested to give better performance than RGB, YCbCr, and HSI models and is successful in the secure transmission of key frames during WCE. Thus, the scope of color conversion is immense in endoscopic images and detail study of which helps in arriving at the appropriate one for a particular application.

20.2 Color Models

The selection of the suitable color model for a particular application depends on the properties of the image and nature of the application. The color models may be classified under three main categories based on the application: device dependent (affected by the signal of the device), user dependent (enable the perception of color), and device independent (not affected by the given device). Another classification would be either psychological spaces or geometric spaces. The psychological spaces arrange the colors based on how they are perceived, whereas the geometric spaces are based on the mathematical layout of the colors. In this section, we discuss the color models suitable for endoscopic images with its components and transform conversion from RGB color space.

20.2.1 RGB Color Models

RGB color model is an additive color model based on Cartesian coordinate system represented by three primary color components, namely, red, green, and blue with its values ranging from 0 to 255. This device-dependent color model has been widely used for endoscopic images as the input image from the image sensors is mostly in RGB format while device-independent models have been derived by deformations of RGB space. The detection and diagnosis performed on endoscopic images sometimes demand the clinical and pathological features of tissues which are not adequately represented in the RGB model. In addition, RGB is not very effective in terms of the bandwidth requirement for each component. The perceptual nonuniformity along with high correlation among its components is a severe drawback which leads to conversion from the RGB domain to other color models.

20.2.2 YCbCr Color Models

YCbCr color model is a luminance-based representation used for digital videos. The three components that are significant in this model are Y representing luminance while red difference Cr and blue difference Cb representing the chrominance components. The RGB to YCbCr conversion formula is given as follows [8]. This model is highly recommended for hardware implementation.

$$Y = 0.299R + 0.587G + 0.114B \tag{20.1}$$

$$Cb = 0.1687R - 0.3313G - 0.5B + 128 \tag{20.2}$$

$$Cr = 0.5R - 0.4187G + 0.0813B + 128 \tag{20.3}$$

20.2.3 YIQ Color Models

YIQ color model [18] is another luminance-based color model that converts RGB inputs from the camera to a luminance (Y) and two chrominance components, namely, I and Q using conversion formula given by (20.4) and (20.5). The luminance component is calculated using (20.1) similar to YCbCr model. The I (orange-blue) and Q (purple-green) indicate the in-phase and quadrature components, respectively. In ref. [12], the focal fibrotic ulcer is detected on the endoscopic image where the larger value of Q indicates the presence of pathology. This is a clinically valid diagnosis and highlights the suitability of this model for abnormality detection.

$$I = 0.596R - 0.274G - 0.322B \tag{20.4}$$

$$Q = 0.211R - 0.523G + 0.312B \tag{20.5}$$

20.2.4 YUV Color Models

The YUV model is specifically used in compression scenario as component values in YUV plane components do not vary much compared to the RGB model. The RGB to YUV converters are present in some image sensors used for capsule endoscopy as it requires only a few shifts and additions. Therefore, this conversion is highly recommended for the implementation inside the capsule with low power and area constraints. In YUV color space, Y is the luminance component, U represents red, and V represents the blue component [4] as given by (20.1), (20.6), and (20.7), respectively. This model is recommended based on the observation that green is significantly absent in such images.

$$U = -0.14713R - 0.28886G - 0.436B \tag{20.6}$$

$$V = 0.615R - 0.51499G - 0.10001B \tag{20.7}$$

20.2.5 YEF Color Models

The YEF color space is yet another model where Y represents luminance, E the difference between luminance and green, and F indicates the difference between luminance and blue component [6] as given in (20.8)–(20.10). The real motivation behind this representation is the dominance of red component and significant absence of green and blue components in the endoscopic images which may not be uniform for all classes of endoscopic images. However, this representation proves itself to be the best candidate for hardware implementation in vivo trials [19].

$$Y = (R/4) + (G/2) + (B/4) \tag{20.8}$$

$$E = (R/8) - (G/4) + (B/8) + 128 \tag{20.9}$$

$$F = (R/8) + (G/8) - (B/4) + 128 \tag{20.10}$$

20.2.6 YCgCo Color Models

Due to the WCE power limits, RGB to YCgCo conversion is employed in ref. [11] on the endoscopic images. The luminance component Y is obtained in a similar way as that of YEF conversion. The formula for the green chroma Cg and orange chroma Co are given in (20.11) and (20.12).

$$Cg = (G/2) - (G/4) - (B/4) \tag{20.11}$$

$$Co = (R/2) - (B/2) \tag{20.12}$$

20.2.7 HSV Color Models

The parameter of brightness present in color model may not be very effective while examining the images under illumination in capsule endoscopy. HSV color model is a cylindrical coordinate system based on hue (H), saturation (S), and value (V) capable of reducing this dependency on brightness as represented by (20.13)–(20.15) [20]. Hue indicates the basic color and is the angle around the color hexagon from a reference point to that particular color. Saturation is considered as the departure of the color from white and value is referred to as the departure of the color from black. Thus, the phenomenal color space model is based on color descriptions rather than individual components and thus, it helps the user to specify the desired color much more easily. HSV space is essentially a device-dependent model perceptually uniform and employed in the way the human eye perceives and interprets the color.

$$H = \begin{cases} 0° & \Delta = 0 \\ 60 \times \left(\frac{G'-B'}{\Delta} \bmod 6\right) & K_{max} = R' \\ 60 \times \left(\frac{B'-R'}{\Delta} + 2\right) & K_{max} = G' \\ 60 \times \left(\frac{B'-R'}{\Delta} + 2\right) & K_{max} = B' \end{cases} \tag{20.13}$$

$$S = \begin{cases} 0 & K_{max} = 0 \\ \frac{\Delta}{K_{max}} & K_{max} \neq 0 \end{cases} \tag{20.14}$$

$$V = K_{max} \tag{20.15}$$

where $K_{min} = min(R', G', B')$, $K_{max} = max(R', G', B')$, and $\Delta = K_{max} - K_{min}$.

20.2.8 HSI Color Models

The HSI color model is similar to the HSV color space that decouples the brightness component from the color carrying components. Thus, the HSI model proves itself to be an ideal tool for developing image processing algorithms based on color perception based on human vision. In this representation, H denotes hue, S saturation, and I intensity given as follows (20.16)–(20.18), where K_{min} is same as that described for HSV model.

$$I = (R + G + B)/3 \tag{20.16}$$

$$H = \cos\left(\frac{(R - G) + (R - B)}{\sqrt[2]{(R - G)^2 + (R - B)(G - B)}}\right) \tag{20.17}$$

$$S = 1 - 3\frac{K_{min}}{I} \tag{20.18}$$

20.2.9 CIE Lab Color Models

The color space defined by the CIE is based on one channel for luminance L and two color channels, a and b. The L component increases from the bottom to the top of the three-dimensional model. The a-axis extends from −a to +a corresponding to green and red, respectively. Similar to this, b-axis extends from −b to +b indicating blue and yellow, respectively. This color space is better suited for many digital image manipulations than the RGB space. The components are expressed by Equations (20.19)–(20.21) taken from ref. [16].

$$L = 116\left(\sqrt[3]{\left(\frac{Y}{Y_0}\right)}\right) - 16 \tag{20.19}$$

$$a = 500\left(\sqrt[3]{\frac{X}{X_0}} - \sqrt[3]{\frac{Y}{Y_0}}\right) \tag{20.20}$$

$$b = 200 \left(\sqrt[3]{\frac{Y}{Y_0}} - \sqrt[3]{\frac{Z}{Z_0}} \right) \tag{20.21}$$

where $X = 0.490R + 0.310G + 200Y = 0.177R + 0.813G + 011B$, $Z = 0.010G + 0.99B$ and X_0, Y_0, Z_0 are X, Y, Z values of white, respectively.

20.3 Performance Measures

20.3.1 Endoscopic Dataset

For experimentation, we have selected WEO Clinical Endoscopy Atlas [21] dataset which contains 147 images classified under various categories, namely, lumen, content, mucosa, flat lesion, protruding lesion, and excavated lesion. It is open access and contains a variety of clinically significant images available. Lumen is the opening inside a tubular body, the presence of foreign bodies or food particles are included under contents category. Images under mucosa indicate Barrett's esophagus and pseudo membranes related with small bowel. Flat lesions are recognized as more difficult to detect while the images of protruding lesions include polyps, tumor/mass, etc., of which malignant lesions are relevant for early gastric cancer. The final category is the excavated lesions mainly including erosions; the most images related to such lesions in the small intestine are connected with bleeding.

20.3.2 Statistical Analysis

The following parameters are evaluated for each color component to better understand the distribution of different components in the color models.

20.3.2.1 Standard Deviation

The standard deviation is a scalar calculated from each color component matrix given as follows [22]:

$$\sigma = \sqrt[2]{\frac{\sum (x_i - \mu)^2}{N}} \tag{20.22}$$

where σ is the standard deviation, x_i is the component value at each pixel, μ is the mean, and N is the number of pixels present in one plane. The square of the standard deviation will give variance.

20.3.2.2 Entropy

The next important parameter is entropy which can be used to characterize the texture of the color image. The entropy or average information of an image can be determined from the image using (20.23) where p contains the histogram counts [23].

$$\text{Entropy} = -\sum p \times \log_2 p \qquad (20.23)$$

20.3.2.3 Skewness

In addition, we have also evaluated the skewness, the third standardized moment of the endoscopic dataset for better analyzing the images [22] as in (20.24). The skewness is the measure of symmetry and describes how pixel values are distributed. The darker pixels tend to be positively skewed than the lighter ones in an image.

$$\text{Skewness} = \frac{\sum_{i=1}^{N} (X_i - \mu)^3 / N}{\sigma^3} \qquad (20.24)$$

20.3.2.4 Kurtosis

Another parameter used for analysis of endoscopic dataset is kurtosis. The kurtosis measures the degree of the peak in the distribution of pixel values. A positive value indicates the values are uniformly distributed and a negative kurtosis indicates the presence of spikes in the distribution. The moment raised to a fourth power [22] is used for computing kurtosis given by (20.25).

$$\text{Kurtosis} = \frac{\sum_{i=1}^{N} (X_i - \mu)^4 / N}{\sigma^4} \qquad (20.25)$$

20.3.2.5 Average Distance

The average distance between the total pixels and maximum primary colors for each component plane is evaluated using (20.26). For an M × N image, the average distance (d_{avg}) between the maximum value of that component and total pixels is calculated [24]. For instance, in the RGB model the average distance for the red component (Rd_{avg}) evaluation, the maximum of R (R_{max}) in the image and R (i, j) indicate the value of red component at (i, j) position are used to compute $R_{d_{avg}}$. Similarly, the evaluation of the green component and blue component is computed by finding the maximum green and blue component values, respectively.

$$R_{d_{avg}} = \left(\frac{1}{M \times N}\right) \sum_{i=0}^{M-1} \sum_{j=0}^{N-1} \left[1 - \frac{R(i,j)}{R_{max}}\right] \qquad (20.26)$$

20.3.2.6 Variance from Average Distance

The variance can be obtained from the average distance [24] for the red component using the formula given in (20.27) where $R_{d_{avg}}$ is obtained from (20.26). The variance is computed similarly for all other components using the same formula replacing R(i, j), R_{max}, and $R_{d_{avg}}$ with their corresponding component values.

$$R_{var} = \left(\frac{1}{M \times N}\right) \sum_{i=0}^{M-1} \sum_{j=0}^{N-1} \left[1 - \frac{R(i,j)}{R_{max}}\right]^2 - R_{d_{avg}}{}^2 \qquad (20.27)$$

20.4 Results and Discussions

In this work, we have extracted the color components present in all the above-discussed color models on the experimental dataset. The component values separated for a single image (image no. 32) are demonstrated in Figure 20.2. It is noticed that in each color model, three components are combined together to form the original image. Figure 20.3 shows the histogram obtained for each color plane. The histogram in addition to the statistical parameter is an ideal tool to study and analyze the endoscopic image. As explicit from the histogram, in the RGB plane, all the three component values are spread from 0 to 255. In YUV representation, U component occupies a narrow region compared

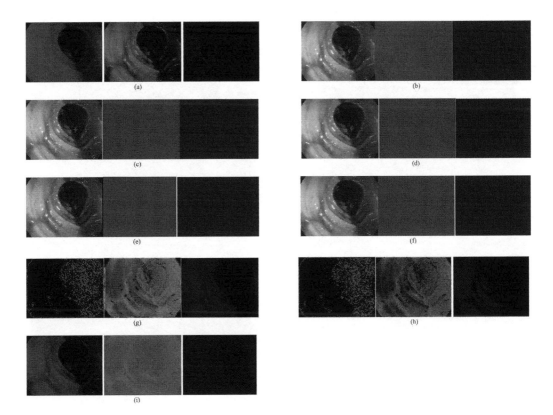

FIGURE 20.2
The three component values representation of each color model: (a) RGB, (b) YCbCr, (c) YIQ, (d) YUV, (e) YEF, (f) YCgCo, (g) HSV, (h) HSI, (i) CIE Lab. (Source: Simulated in MATLAB 2017).

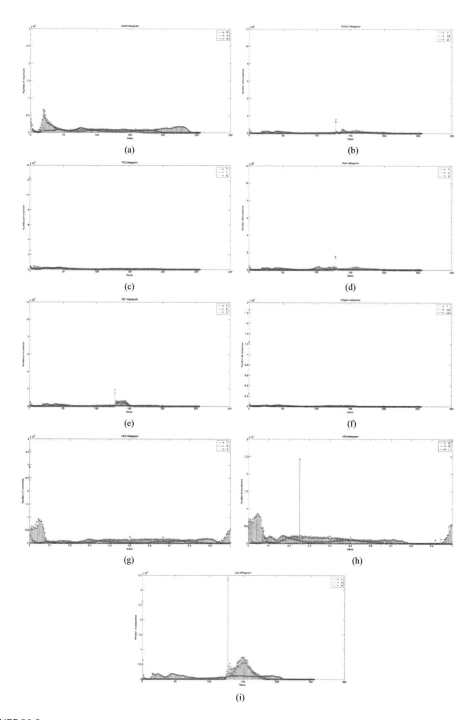

FIGURE 20.3
The histogram of each color model: (a) RGB, (b) YCbCr, (c) YIQ, (d) YUV, (e) YEF, (f) YCgCo, (g) HSV, (h) HSI, (i) CIE Lab. (Source: Simulated in MATLAB 2017).

to Y and V, while in YEF and YCgCo, E and Cg components, respectively, occupy a narrow histogram. On the other hand, in YCbCr color space the component values are spread only for the luminance component while for Cb and Cr, it is narrow indicating its superiority in compression application. For HSI and HSV models, like RGB model the histogram is spread in all three component values as indicated in Figures 20.2(g) and (h), respectively; however, the value range is 0 to 1. It should be noted that these histograms are based on a single image in the dataset. To investigate further, we have taken several statistical measures discussed in the previous section on our dataset. Table 20.1 consolidates the results obtained for various parameters on each color component when simulated for the entire endoscopic database. As there are varieties of images available in the endoscopic dataset under study, this table put forward the exact pattern in which component values change in each color plane.

The standard deviation is one of the best statistical parameters which helps to determine and analyze the component values present in a color model. The results

TABLE 20.1

Results obtained for performance measures

Color space	Components	Standard deviation	Entropy (bits/pixel)	Skewness	Kurtosis	Average distance	Variance from average distance
RGB	R	75.13	0.5	−0.22	2.08	0.56	0.09
	G	52.59	0.5	0.36	2.88	0.71	0.05
	B	42.5	0.52	1.05	6.22	0.78	0.03
YCbCr	Y	57.22	0.56	0.09	2.36	0.69	0.05
	Cb	11.89	0	0.03	2.79	1.61	0.45
	Cr	16.31	0	−0.24	9.61	0.64	0.08
YIQ	Y	57.22	0.56	0.09	2.36	0.67	0.05
	I	22.11	0.89	−0.16	2.68	4.1	8.35
	Q	4.9	1.27	−0.1	6.65	0.65	0.07
YUV	Y	57.22	0.56	0.09	2.36	0.69	0.05
	U	10.36	0.34	0.03	2.79	0.08	0
	V	20.06	0.92	−0.24	9.61	0.12	0
YEF	Y	54.62	0.54	0.16	2.51	0.67	0.05
	E	2.95	0	0.05	7.03	0.16	0.01
	F	7.09	0	−0.1	2.63	0.19	0.01
YCgCo	Y	54.62	0.54	0.16	2.51	0.67	0.05
	Cg	5.9	0.56	−0.05	7.03	0.64	8.35
	Co	21.5	0.82	−0.16	2.51	0.89	0.06
HSV	H	0.16	4.18	−0.22	2.08	0.89	0.03
	S	0.25	5.98	0.36	2.88	0.56	0.07
	V	0.29	6.58	1.05	6.22	0.56	0.09
HSI	H	0.17	4.31	−0.22	6.22	0.86	0.04
	S	0.25	5.9	0.36	2.88	0.55	0.07
	I	0.22	6.4	1.05	6.22	0.68	0.05
CIE Lab	L	28.22	0.49	−0.73	2.66	0.46	0.08
	a	8.45	0.98	−0.88	37.88	0.89	0.01
	b	10.86	0.86	1.18	16.6	0.85	0.02

show that it follows the same conclusions which we arrived previously based on histogram analysis. In RGB, HSI, and HSV color spaces, the average standard deviation values are uniformly distributed in all the three component values implying all have equal weightage in determining the image. On the other hand, in all other luminance-based color models, Y component shows superior values for standard deviation than the chrominance components. As the luminance component plays a vital role in determining the shape and structure of an image, preserving this component intact would help greatly to convey the significant information for diagnosis purpose. Moreover, due to the limitations of human visual system in perceiving the colors, the chrominance components are less significant and this opens up the possibility to reduce the chrominance data to a great extent particularly applications related to compression. The energy content of each component is described by the entropy value. It is noticed that the average entropy value of chrominance component is 0 in YCbCr and YEF color spaces. Thus, much energy compaction is possible in such representation which leads to its increased acceptability for subsampling the chrominance data effectively without visible artifacts. On the contrary, in YIQ color space, the Q component carries much information and its value plays an important role in detecting abnormality as discussed in ref. [12]. In three models, RGB, HSI, and HSV, the information content is uniformly distributed and this leads to an important conclusion regarding its unsuitability for compression. It can be seen from the table that the skewness for component values can either be positive or negative value. Note that the endoscopic procedure is performed with various illumination conditions and therefore, based on the light intensity the dark and light regions are possible in the image region. The positive value indicates darker regions in the image while a negative value shows lighter image regions. Like skewness, average kurtosis is also used to analyze the distribution of image components. The presence of positive values indicates that the components are uniformly distributed; this is mainly due to the homogeneity of color distribution and the absence of sharp edges in such images.

Another statistical parameter is the average distance measured based on the maximum value component in each plane; its value helps to interpret the significant component in a color space based on its occurrence. For instance, in RGB color space, R component gives the shortest distance indicating it occupies a large proportion of the image whereas the blue component is the least and smallest proportion. Similarly, the application of YCbCr color model on the endoscopic dataset indicates the Cb (luminance minus blue) component occupies more proportion of the image as blue difference signal indicates the dominance of red and green values indirectly. The final parameter used is variance evaluated from the average distance which also leads to a similar interpretation. In RGB color space, the B component has a minimum value and R has maximum value implying red component must be preserved carefully than the other two components. To be more precise, the red component is the dominant one which conveys the information. It should be noted that the dynamic range of red is broader than the other two components.

20.5 Conclusion

The color models are representation involving conversion from basic RGB model to several standard predefined models so as to help further proceeding on image data. In this chapter we mainly focus on studying the endoscopic images and implementing various color conversions on selected widely used nine color models. In order to strengthen our study, we have performed various statistical measures to analyze the experimental dataset. In addition, the study of the histogram of various color models aids in the discussion of a model from an application perspective. It has been concluded that each color model is used in different applications depending upon its components and its purpose is to facilitate the specification of colors in some standard generally accepted way. The models discussed here are confined to image display on the specific hardware platform. Thus, this work opens up a detailed analysis of endoscopic images aiming to help researchers to simplify their work by choosing an appropriate color model suitable for their purpose.

References

[1] G. Ciuti, A. Menciassi, and P. Dario, "Capsule endoscopy : From current achievements to open challenges," *IEEE Trans. Biomed. Eng.*, vol. 4, pp. 59–72, 2011.

[2] S. A. Karkanis, D. K. Iakovidis, D. E. Maroulis, D. A. Karras, and M. Tzivras, "Computer-aided tumor detection in endoscopic video using color wavelet features," *IEEE Trans. Inf. Technol. Biomed.*, vol. 7, no. 3, pp. 141–152, 2003.

[3] B. Münzer, K. Schoeffmann, and L. Böszörmenyi, "Content-based processing and analysis of endoscopic in content-based processing and analysis of endoscopic images and videos: A survey," *Multimed. Tools Appl.*, vol. 77, no. 1, pp. 1323–1362, 2018.

[4] J. Wu, and Y. Li, "Low-complexity video compression for capsule endoscope based on compressed sensing theory," *Proc. 31st Annu. Int. Conf. IEEE Eng. Med. Biol. Soc. Eng. Futur. Biomed. EMBC 2009*, pp. 3727–3730, 2009.

[5] T. H. Khan, S. Member, and K. A. Wahid, "Low power and low complexity compressor for video capsule endoscopy," *1542 IEEE Trans. Circuits Syst. Video Technol.*, vol. 21, no. 10, pp. 1534–1546, 2011.

[6] T. H. Khan, and K. A. Wahid, "White and narrow band image compressor based on a new color space for capsule endoscopy," *Signal Process. Image Commun.*, vol. 29, no. 3, pp. 345–360, 2014.

[7] J. Thoné, J. Verlinden, and R. Puers, "An efficient hardware-optimized compression algorithm for wireless capsule endoscopy image transmission," *Proc. Eng.*, vol. 5, pp. 208–211, 2010.

[8] D. Turgis, and R. Puers, "Image compression in video radio transmission for capsule endoscopy," *Sens. Actuators, A Phys.*, vol. 123–124, no. April, pp. 129–136, 2005.

[9] M. S. Imtiaz, and K. A. Wahid, "Color enhancement in endoscopic images using adaptive sigmoid function and space-variant color reproduction," *Comput. Math. Methods Med.*, 2015.

[10] Y. Yuan, B. Li, and M. Q. H. Meng, "Bleeding frame and region detection in the wireless capsule endoscopy video," *IEEE J. Biomed. Heal. Informatics*, vol. 20, no. 2, pp. 624–630, 2016.

[11] P. Turcza, and M. Duplaga, "Low-power image compression for wireless capsule endoscopy," *2007 IEEE Int. Work. Imaging Syst. Tech.*, pp. 1–4

[12] P. Szczypiński, A. Klepaczko, M. Pazurek, and P. Daniel, "Texture and color based image segmentation and pathology detection in capsule endoscopy videos," *Comput. Methods Programs Biomed.*, vol. 113, no. 1, pp. 396–411, 2014.

[13] D. K. Iakovidis, D. E. Maroulis, and S. A. Karkanis, "An intelligent system for automatic detection of gastrointestinal adenomas in video endoscopy," *Comput. Biol. Med.*, vol. 36, no. 10, pp. 1084–1103, 2006.

[14] B. V. Dhandra, R. Hegadi, M. Hangarge, and V. S. Malemath, "Analysis of abnormality in endoscopic images using combined HSI color space and watershed segmentation," *18th Int. Conf. Pattern Recognit.*, vol. 4, pp. 18–21, 2006.

[15] M. Mackiewicz, J. Berens, and M. Fisher, "Wireless capsule endoscopy color video segmentation," *IEEE Trans. Med. Imaging*, vol. 27, no. 12, pp. 1769–1781, 2008.

[16] M. P. Tjoa, S. M. Krishnan, C. Kugean, P. Wang, and R. Doraiswami, "Segmentation of clinical endoscopic image based on homogeneity and hue," *2001 Conference Proceedings of the 23rd Annual International Conference of the IEEE Engineering in Medicine and Biology Society*, vol. 3. pp. 2665–2668, 2001.

[17] K. Muhammad, J. Ahmad, M. Sajjad, S. Rho, and S. W. Baik, "Evaluating the suitability of color spaces for image steganography and its application in wireless capsule endoscopy," *2016 International Conference on Platform Technology and Service (PlatCon)*, Jeju, pp.1–3, 2016.

[18] S. H. Kwok, "Efficient gamut clipping for color image processing using LHS and YIQ," *Opt. Eng.*, vol. 42, no. 3, pp. 701–711, 2003.

[19] T. H. Khan, and K. A. Wahid, "Design of a lossless image compression system for video capsule endoscopy and its performance in in-vivo trials," *Sensors (Switzerland)*, vol. 14, no. 11, pp. 20779–20799, 2014.

[20] D. Khattab, H. M. Ebied, A. S. Hussein, and M. F. Tolba, "Color image segmentation based on different color space models using automatic GrabCut," *Sci. World J.*, vol. 2014, 2014.

[21] WCE Clinical Endoscopy Atlas Available online: www.endoatlas.org/

[22] C. C. Hsia, Y. W. Hung, Y. H. Chiu, and C. H. Kang, "Bayesian classification for bed posture detection based on kurtosis and skewness estimation," *2008 10th IEEE Intl. Conf. e-Health Networking, Appl. Serv. Heal.*, pp. 165–168, 2008.

[23] A. Lesne, "Shannon entropy: A rigorous notion at the crossroads between probability, information theory, dynamical systems and statistical physics," *Math. Struct. Comput. Sci.*, vol. 24, no. 3, 2014.

[24] M. C. Lin, and L. R. Dung, "A subsample-based low-power image compressor for capsule gastrointestinal endoscopy," *EURASIP J. Adv. Signal Process.*, 2011, 2011.

21

Adaptive Fractal Image Coding Using Differential Scheme for Compressing Medical Images

P. Chitra, M. Mary Shanthi Rani, and V. Sivakumar

The Gandhigram Rural Institute (Deemed to be University), Dindigul, Tamil Nadu, India
Asia Pacific University of Technology and Innovation, Kuala Lumpur, Malaysia

21.1 Introduction

Digital image occupies voluminous data storage for storing and transmitting data. Image compression plays a vital role for reducing the storage space by eliminating redundant and irrelevant storage space [1-3]. It also represents an image with a reduced bit storage. Image compression is widely classified into lossy and lossless compression techniques. Lossless compression technique perfectly reconstructs the image without losing any information. Most of the research problems prefer lossless technique for the reason of image quality. Despite the importance of image quality in medical field, compression is essential for transmitting voluminous image data [4-6]. So, lossless image compression is the preferred method for medical image compression as high quality is required for accurate diagnosis. In lossy image compression process, the image will be represented by eliminating the redundant and irrelevant information from the original data. Generally, lossy image compression is a best choice for photographic/still images, as the human vision could not identify the minor changes in the image visually. Fractal coding is adopted for lossy compression process [7-9].

21.1.1 Fractal Image Coding

Generally, fractals are considered as geometric and infinite patterns which are self-similar across different scales. The fractal is created by using recursion with the simple iterative process. The number of iterations of the fractal process determines the strength of the computation complexity. Fractals are represented by using images of dynamic systems with irregular shapes. Fractal patterns are extremely familiar such as trees, rivers, coastlines, mountains, clouds, seashells, hurricanes, etc.

The core of fractal image compression is to generate a fractal which is very similar to other parts of an image. The basic process of fractal image compression divides an image into range blocks and domain blocks. The range blocks represent the original image blocks and the domain blocks are created by using range blocks of the image.

21.2 Related Work

Fractal image coding is a popular technique and is chosen widely for many applications. There are various algorithms proposed using fractal image coding. Gupta et al. (2017) [10] described a comparative analysis of edge-based fractal image compression using nearest neighbor technique in various frequency domains. Ge et al. (2006) [11] proposed a box-counting method for calculating fractal dimension of urban forms based on remote sensing imaging. Bhavani and Thanushkodi (2013) [12] described the performance of fractal-based coding algorithms such as standard fractal coding, quasi-lossless fractal coding, and improved quasi-lossless fractal coding. Muruganandham and Banu (2010) [13] proposed a fast fractal encoding system using particle swarm optimization. Truong et al. (2000) [14] proposed a novel encoding algorithm for fractal image compression. The author proves that the algorithm reduces computation time and redundant storage as well. Maha Lakshmi (2016) [15] proposed a fractal image compression using neural networks for encoding specifically MRI images. The trained net is used for encoding other images eventually reducing the encoding time. Xing-yuan et al. (2009) [16] proposed an efficient method of fractal image compression based on spatial correlation and hybrid genetic algorithm. Zheng (2006) [17] investigated an improved method of generating a binary image affine iterated function system (IFS) by using genetic algorithm. Zheng (2013) [18] examines the affine parameters in IFS with different image quality measurements. This method proves that the positive correlation exists between the image contrast of fractal decoded image and affine scalar multiplier.

21.3 Fractal Coding Using Differential Scheme

The proposed work presents a novel and simple fractal image coding using differential scheme for compressing medical images. The proposed method has been implemented in two phases. In the first phase, the input image is converted into a set of image blocks, and vector quantization (VQ) process is applied to the image blocks using fractal dimension scheme. The novelty is that multistage VQ is applied to reduce the dimensions of fractals. Differential coding is applied for improving the quality in the second stage. Finally, the image has been reconstructed using fractal decoding process.

Initially, the input image is divided into non-overlapped [4, 4] blocks which are the key elements for generating reference blocks. The reference blocks also known as code vectors in VQ terminology are generated by using K-means clustering algorithm, an effective technique to generate a primary codebook. This primary codebook is further divided into [2, 2] blocks and is used as input for the second stage of VQ to reduce its size further. In stage 2 VQ, a second codebook has been created by reducing the dimensions and the size of the primary codebook. This is considered as the fractal codebook and each code vector is considered as the fractal, that is, the repeating patterns in the input image. Finally, the third stage of VQ generates the master fractal (MF) codebook from the original image blocks with fractal codebook as the initial seeds for the clusters. The codebook could be created in a desired size where a higher codebook yields good image quality.

In the second phase of proposed work, differential coding has been implemented using reconstructed fractal image (F) and original image (O). Residual vectors are computed by finding difference between original image vector and fractal image vector.

$$RV(i, :) = O(i, :) - F(i, :) \tag{21.1}$$

The residual values are normalized to positive integers to preserve the sign, by adding the maximum residual value NV.

$$NRV(i, :) = RV(i, :) + NV \tag{21.2}$$

where NRV represents the normalized residual vector. The normalized vector has been created for preserving the quality of image blocks.

These normalized residual vectors are classified into low variant and high variant vectors based on variance threshold. This variance threshold is user specified and can be used to tune the quality of the reconstructed image. The third stage applies adaptive quantization to the residual vectors belonging to two classes. High variant vectors are subjected to vector quantization generating a primary residual (PR) codebook. Scalar quantization is applied to low variant vectors by mapping each vector to its mean value forming the secondary residual (SR) codebook. Moreover, this low variant codebook is constructed by taking only the unique mean values.

However, the compressed stream will have MF codebook, renormalized PR codebook, and SR codebook along with their respective code indices.

During the decoding process, the residual vectors are reconstructed using primary and secondary residual codebooks first and are added to the vectors reconstructed using master fractal codebook to generate the decoded image, that is, the final image vector RI_i RI_i is reconstructed by adding the matching code vectors for each input vector from the three codebooks using its respective code index say "i."

$$RI(i, :) = MF(i, :) + PR(i, :) + SR(i, :) \tag{21.3}$$

21.3.1 Proposed Algorithm

Phase I

- Divide the input image into non-overlapped blocks. Each block of image is treated as a vector.
- Generate the initial codebook by using k-means clustering with image blocks in Stage I VQ.
- Divide the codebook of Stage I VQ into [2, 2] sub-blocks and it is used as a primary codebook for Stage II VQ.
- Generate the secondary codebook by reducing the dimensions and size of the primary codebook in Stage II VQ.
- Generate the master fractal codebook from the original image blocks with secondary codebook as the initial seeds.

Phase II

- Calculate the residual between the fractal reconstructed image blocks and the original image blocks using Eq. (21.1).
- Find the mean value for each residual vector.

- Classify the residual vectors into high variant and low variant vectors based on a variance threshold.
- Normalize the residual vector by adding the maximum residual value NV using Eq. (21.2).
- Construct PR codebook by applying VQ to the high variant residual image blocks.
- Construct SR codebook by storing unique mean values of low variant blocks.

The compressed stream consists of the MF, PR, and SR codebooks.

Decoding Process

- Find the matching residual code vectors from PR and SR using the code indices.
- Renormalize the residual code vectors.
- Reconstruct the image by adding the renormalized residual code vectors to the respective reconstructed MF image blocks using Eq. (21.3).

The process of proposed algorithm is visually represented in Figure 21.1 and Figure 21.2.

Phase I

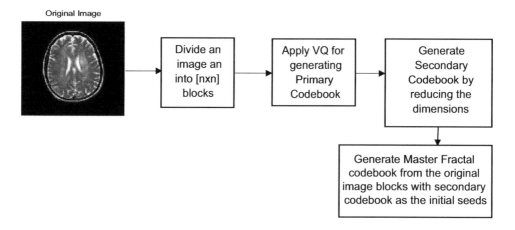

Phase II

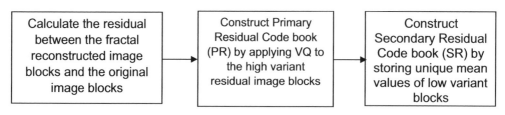

FIGURE 21.1
Encoding Process of the Proposed Method

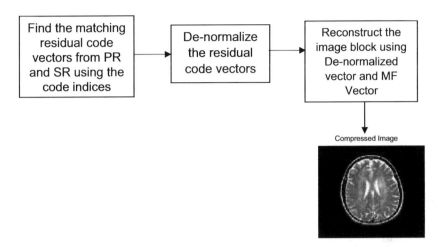

FIGURE 21.2
Decoding Process of the Proposed Method.

21.4 Results and Discussion

The performance of the proposed method has been evaluated using the standard evaluation metrics in terms of quality and compression efficiency. The experiments of the proposed method are conducted using four different medical image modalities MRI_T2 weighted image (RIDER NEURO MRI), Mammogram Image (VICTRE), CT Colon (CT Colonography), and Image and X-ray (TCGA-LUAD) Image [20-23]. The image quality achieved by the proposed method has been assessed using peak signal to noise ratio (PSNR) calculated based on mean square error (MSE) and structural similarity (SSIM) metric. The compression efficiency of the proposed method has been measured using compression ratio (CR) and bit rate (BR) using bit per pixel (bpp). The computation time has been calculated for assessing the computational efficiency of the proposed algorithm.

Table 21.1 shows the performance of the proposed method using different codebook sizes. The performance of each image varies according to their metrics.

Table 21.1 demonstrates the effectiveness of the proposed method by achieving higher PSNR and CR values using [2, 2] block size. Also, it minimizes the bit rate of the compressed image. It is obvious that the proposed method has achieved PSNR value in the considerable range (47–55) which is optimal for medical image diagnosis and SSIM closer to 1 exhibiting near-lossless quality achievement by the proposed method.

Table 21.2 examines the comparative analysis of the performance of proposed method using similar existing method.

Table 21.2 shows that the proposed method outperforms similar existing method by achieving higher PSNR and CR value. The comparative analysis of the performance of proposed method is graphically represented in Figure 21.3 using PSNR and CR values.

TABLE 21.1

Performance Analysis of the Proposed Method

Image	Codebook	Quality Metrics			Compression Efficiency		
		MSE	PSNR	SSIM	CR	BR	CT
MRI_T2 Brain	256	0.561	50.636	0.806	13.689	0.584	12.010
	512	0.281	53.632	0.833	10.675	0.749	16.055
	1024	0.184	55.469	0.878	8.743	0.914	10.487
Mammogram	256	0.253	54.086	0.961	24.021	0.333	6.592
	512	0.217	54.764	0.953	17.998	0.444	6.804
	1024	0.119	57.360	0.953	12.919	0.619	11.738
CT_Colon	256	1.166	47.460	0.844	11.538	0.693	12.368
	512	1.032	47.992	0.886	9.656	0.828	10.129
	1024	0.544	50.774	0.938	7.906	1.011	11.075
X-ray	256	1.164	47.469	0.991	19.060	0.419	8.878
	512	0.953	48.337	0.990	13.552	0.590	10.424
	1024	0.465	51.454	0.991	10.496	0.762	7.711

TABLE 21.2

Comparative Analysis between the Proposed Method and Similar Existing Method

Image	Methods	PSNR	CR	ET
MRI_Brain	Existing Method[15]	39.91	12.87	503.25
	Proposed Method	50.63	13.68	12.01

TABLE 21.3

Comparative Analysis between the Proposed Method and Existing Methods for Other Medical Images

Images 512x512 pixels	Algorithms	bit per pixel (bpp)			
		0.19		0.25	
		PSNR	MSE	PSNR	MSE
	CLC[19]	26.78	136.48	28.05	101.88
MR Cardiac	CLC+SEC[19]	35.10	20.10	35.27	19.32
	Proposed Method	56.96	0.13	58.32	0.09
	CLC[19]	28.19	98.65	28.82	85.33
MRI Skull	CLC+SEC[19]	35.80	17.10	36.08	16.04
	Proposed Method	48.64	0.88	51.14	0.49
	CLC[19]	25.01	205.15	26.37	150.00
Ultrasound Liver Cyst	CLC+SEC[19]	30.54	57.42	31.07	50.83
	Proposed Method	44.31	2.40	45.07	2.02

Table 21.3 shows the performance of the comparative analysis of the performance of the proposed method and existing methods for medical images.

The superior performance of the proposed method is exhibited in Table 21.3 compared to existing methods [19]. Also, it is worthy to note that the proposed work yields higher PSNR range (44–58) over the existing CLC [19] and CLC+SEC [19] methods.

The visual representation of the proposed method is presented in Figures 21.4–21.7, respectively. The outcome of the results has demonstrated the quality of the compressed image.

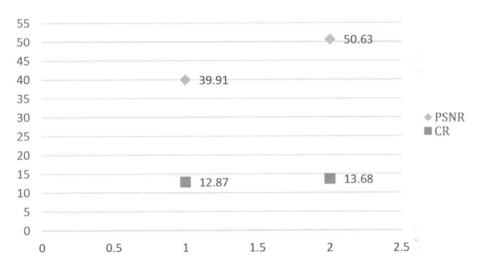

FIGURE 21.3
Graphical Representation of the Proposed Method.

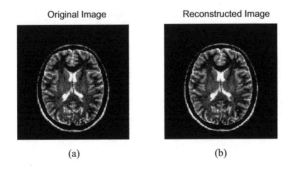

Original Image   Reconstructed Image

(a)   (b)

FIGURE 21.4
(a) Original MRI_Brain Image, (b) Reconstructed MRI_Brain Image Using Proposed Method.

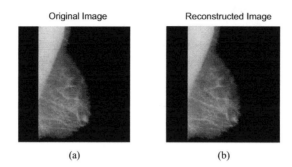

FIGURE 21.5
(a) Original Mammogram Image, (b) Reconstructed Mammogram Image Using Proposed Method.

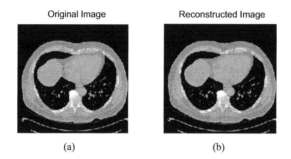

FIGURE 21.6
(a) Original CT Colon Image, (b) Reconstructed CT Colon Image Using Proposed Method.

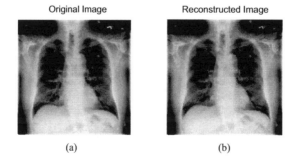

FIGURE 21.7
(a) Original X-ray Image, (b) Reconstructed X-ray Image Using Proposed Method.

21.5 Conclusions

In this research work, a novel and efficient algorithm has been proposed using fractal image coding and differential scheme. The experimental results have proved their efficiency of the proposed algorithm by achieving higher PSNR and CR values. Also, our proposed work has reduced the computation time. This algorithm is well-suited for medical image applications that involve humongous storage and transmission of medical images.

References

[1] K. Somasundaram and M. M. S. Rani, Novel K-Means Algorithm for Compressing Images, *International Journal of Computer Applications*, 18(8), pp. 9–13, 2011.

[2] K. Somasundaram and M. M. S. Rani, Mode Based K-Means Algorithm with Residual Vector Quantization for Compressing Images, International Conference on "Control, Computation and Information Systems" (Springer-Verlag CCIS 140), pp. 105–112, 2011.

[3] M. M. S. Rani, P. Chitra, and R. Vijayalakshmi, Image Compression Based on Vector Quantization Using Novel Genetic Algorithm for Compressing Medical Images, *International Journal of Computer Engineering and Applications*, XII(I), pp. 104–114, Issue, January 2018.

[4] P. Chitra and M. M. S. Rani, Modified Haar Wavelet Based Method for Compressing Medical Images, *International Journal of Engineering and Techniques (IJET)*, 1(2), pp. 243–251, 2018.

[5] M. M. S. Rani, P. Chitra, and K. Mahalakshmi, A Novel Approach of Vector Quantization Using Modified Particle Swarm Optimization Algorithm for Generating Efficient Codebook, *International Journal of Advanced Research in Computer Science*, 8(9), pp. 294–298, 2017.

[6] M. M. S. Rani, "Adaptive Classified Pattern Matching Vector Quantization Approach for Compressing Images," The 2009 International Conference on Image Processing, Computer Vision & Pattern Recognition Proceedings, Las Vegas, USA, pp. 532–538, 2009.

[7] M. M. S. Rani, Residual Vector Quantization Based Iris Image Compression, *International Journal of Computational Intelligence Studies, Inder Science Publishers*, 3(4), pp. 329–334, 2014.

[8] M. M. S. Rani and P. Chitra, Novel Hybrid Method of Haar-Wavelet and Residual Vector Quantization for Compressing Medical Images, *2016 IEEE Conference on Advances in Computer Applications (ICACA)*, 1, pp. 321–326, 2016.

[9] M. M. S. Rani and P. Chitra, A Hybrid Medical Image Coding Method Based on Haar Wavelet Transform and Particle Swarm Optimization Technique, *International Journal of Pure and Applied Mathematics*, 118(8), pp. 3056–3067, 2018.

[10] R. Gupta, D. Mehrotra, and R. K. Tyagi, Comparative Analysis of Edge-Based Fractal Image Compression Using Nearest Neighbor Technique in Various Frequency Domains, *Alexandria Engineering Journal*, 57(3), pp. 1525–1533, 2018.

[11] M. Ge, Q. Lin, and W. Lu, Realizing the Box-counting Method for Calculating Fractal Dimension of Urban Form Based on Remote Sensing Image, *IEEE*, 12(4), pp. 265–270, 2009.

[12] S. Bhavani and K. G. Thanushkodi, Comparison of Fractal Coding Methods for Medical Image Compression, *IET Image Process*, 7(7), pp. 686–693, 2013.

[13] A. Muruganandham and R. S. D. W. Banu, Adaptive Fractal Image Compression Using PSO, ICEBT 2010, *Procedia Computer Science*, 2, pp. 338–344, 2010.

[14] T.-K. Truong, J.-H. Jeng, I. S. Reed, P. C. Lee, and A. Q. Li, A Fast Encoding Algorithm for Fractal Image Compression Using the DCT Inner Product, *IEEE Transactions on Image Processing*, 9(4), pp. 529–535, April 2000.

[15] G. V. Maha Lakshmi, Implementation of Image Compression Using Fractal Image Compression and Neural Networks for MRI Images, *IEEE, 2016 International Conference on Information Science (ICIS)*, pp. 60–64, 2016.

[16] W. Xing-yuan, F.-P. Li, and S.-G. Wang, Fractal Image Compression Based on Spatial Correlation and Hybrid Genetic Algorithm, *Journal of Visual Communication and Image Representation*, 20, pp. 505–510, 2009.

[17] J. Wang, N. Zheng, Y. Liu, and G. Zhou, Parameter Analysis of Fractal Image Compression and Its Applications in Image Sharpening and Smoothing, *Signal Processing: Image Communication*, 28, pp. 681–687, 2013.

[18] Y. Zheng, An Improved Fractal Image Compression Approach by Using Iterated Function System and Genetic Algorithm, *International Journal of Computers and Mathematics with Applications*, Elsevier, Vol. 51, pp. 1727–1740, 2006.

[19] T. Phanprasit, K. Hamamoto, M. Sangworasil, and C. Pintavirooj, Medical Image Compression Using Vector Quantization and System Error Compression, *IEEJ Transactions on Electrical and Electronic Engineering*, 10(5), pp. 554–566, 2015.

[20] K. Clark, B. Vendt, K. Smith, J. Freymann, J. Kirby, P. Koppel, S. Moore, S. Phillips, D. Maffitt, M. Pringle, L. Tarbox, and F. Prior, The Cancer Imaging Archive (TCIA): Maintaining and Operating a Public Information Repository, *Journal of Digital Imaging*, 26 (6), pp. 1045–1057, 2013.

[21] J. E. E. Oliveira, M. O. Gueld, A. D. A. Araújo, B. Ott, and T. M. Deserno, Towards a Standard Reference Database for Computer-Aided Mammography, *Proceedings of SPIE*, 6915, Paper ID 69151Y, 2008.

[22] https://wiki.cancerimagingarchive.net

[23] www.mammoimage.org/databases

Index

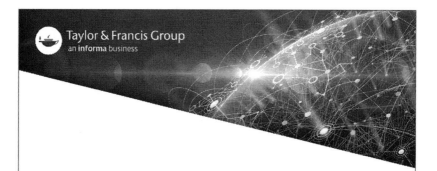